HAZARDOUS MATERIALS INCIDENTS

CHRIS HAWLEY

DELMAR

THOMSON LEARNING ™

Australia Canada Mexico Singapore Spain United Kingdom United States

DELMAR

THOMSON LEARNING

Hazardous Materials Incidents
Chris Hawley

Business Unit Director:
Alar Elken

Executive Editor:
Sandy Clark

Acquisitions Editor:
Mark Huth

Editorial Assistant:
Dawn Daugherty

Executive Marketing Manager:
Maura Theriault

Executive Production Manager:
Mary Ellen Black

Production Manager
Larry Main

Production Editor
Tom Stover

Marketing Coordinator:
Brian McGrath

Cover Designer:
Michael Egan

COPYRIGHT © 2002 by Delmar
a division of Thomson Learning, Inc. Thomson Learning™ is a trademark used herein under license

Printed in Canada
2 3 4 5 6 7 8 9 XXX 05 04 02

For more information contact Delmar,
3 Columbia Circle, PO Box 15015,
Albany, NY 12212-5015.

Or find us on the World Wide Web at http://www.delmar.com or
http://www.autoed.com

For permission to use material from this text or product, contact us by
Tel   (800) 730-2214
Fax   (800) 730-2215
www.thomsonrights.com

ISBN: 0-7668-4296-7

Library of Congress Cataloging-in-Publication Data

NOTICE TO THE READER

Publisher does not warrant or guarantee any of the products described herein or perform any independent analysis in connection with any of the product information contained herein. Publisher does not assume, and expressly disclaims, any obligation to obtain and include information other than that provided to it by the manufacturer.

The reader is notified that this text is an educational tool, not a practice book. Since the law is in constant change, no rule or statement of law in this book should be relied upon for any service to any client. The reader should always refer to standard legal sources for the current rule or law. If legal advice or other expert assistance is required, the services of the appropriate professional should be sought.

The Publisher makes no representation or warranties of any kind, including but not limited to, the warranties of fitness for particular purpose or merchantability, nor are any such representations implied with respect to the material set forth herein, and the publisher takes no responsibility with respect to such material. The publisher shall not be liable for any special, consequential, or exemplary damages resulting, in whole or part, from the readers' use of, or reliance upon, this material.

This text is dedicated to my sons, Timothy and Matthew.
They are the greatest sons a father could ask for.

CONTENTS

FOREWORD

Emergency responders have been going to hazardous materials incidents long before the federal government even invented the term *Hazardous Materials*. In most of these early incidents, which we used to call chemical fires, we rushed into the spill, rescued people, and usually put water on the incident until, somehow, it went away.

In those early days, emergency responders paid a heavy price for their heroic actions. Little concern was given to special tactics. The science of personal protective clothing was archaic and most of the responders truly believed that "real men" did not need masks. This lack of technical information and macho ideals proved disastrous for the American fire service.

We now have established standards that require a greatly improved level of knowledge, which will allow us to do a better and safer job.

Hazardous Material Incidents is intended to be a textbook used by first responder operations classes. These are generally the first emergency responders on the scene. The actions of these first responders are absolutely critical to a successful outcome. Their defensive actions will protect the community, minimize the damage, and ensure a safe working atmosphere.

This book meets and exceeds NFPA 472 at the operations level. If emergency responders understand and follow this book, everyone will benefit.

What makes this text so important is that it gives more than the basic technical information. It shows responders how to do risk-based response. By understanding the management principles of what risks we take for what gains and losses will result, we need no longer charge forward simply out of a sense of duty.

I encourage every reader to study this text and learn well. Your life could depend on it!

John M. Eversole
Special Operations Chief
Chicago Fire Department
Chicago, IL

PREFACE

This text is designed for emergency responders, including fire, EMS, law enforcement, and private industrial responders. This text is adaptable to a variety of disciplines. For example, as a result of growing concerns over terrorism, many responders who were not involved in hazardous materials response are now involved. To function at a potential terrorism event, one needs to have hazardous materials training. One unique feature is that this text incorporates terrorism training.

What I have tried to do in this text is to take what I have learned over the years and incorporate this "street knowledge" into a readable format. Many good training programs exist, but most instructors use a variety of programs in their delivery. This text is a compilation of these various approaches. This text is designed to meet and exceed the objectives outlined in NFPA 472 *Standard for Professional Competence of Responders to Hazardous Materials Incidents.* Following this text students will meet both the hazardous materials awareness level and the hazardous materials operations level. In some cases the material in this text is at the hazardous materials technician level, referred to as OPS-plus. The division of a responder operating at the OPS level is often blurred. Many emergency response organizations have first responders trained to the operations level, but they are assigned to operate air monitoring devices. This text addresses that gap in training by covering those types of responses at the OPS-plus level. This text is designed to be used by students taking a hazardous materials operations class. A companion workbook for students accompanies this text, and an instructor's guide is available as well, both published by Delmar.

The text is built on a logical progression with each chapter building on the next. At the end of each chapter are review questions to enhance learning. The student workbook expands this review process through several questioning techniques. As much as we would like the student to learn everything that is presented in the text, the reality is that this subject is too complex to retain every detail. To reinforce important topics, each safety and caution note is highlighted and separated from the text for easy identification. Key terms are highlighted and defined in the Key Terms section at the end of each chapter; a complete Glossary is also provided.

Each unit has case studies and a number of text boxes that brings real-life experiences to the student. These case studies come from years of experience, and nothing is held back. If mistakes were made, they are discussed; learning from others' mistakes is a good way to learn. Each chapter has a Street Story provided by HAZMAT responders from around the country. A recurrent theme runs through each of them. Each of the stories has a lot of lessons learned, all very personal to that writer. I am very thankful that these professionals, all good friends, have chosen to share with you some personal stories that influenced their lives and their careers. The authors of these Street Stories are the best in the HAZMAT business, but years ago they sat where you are sitting now, starting their careers. Whether you are starting out in a career organization or a volunteer force, I advise you to read each of these stories about experiences in which the authors and their crews could have been seriously injured or killed. When reading the chapter, review in your mind the clues that were present that would have alerted you to the presence of hazardous materials. When you are done with that chapter, go back and reread the Street Story. These true stories offer tremendous potential for learning, and you should take advantage of it.

The learning process involved with hazardous materials is an ever-expanding process. Each day the number of chemicals used by industries and transported across the country is on the increase, and the complexity of our job is increasing as well. Terrorism adds a unique and terrible twist to the emergency response field. The task of responding to terrorism events in most cases has landed to the HAZMAT teams in conjunction with the local police department. If you are involved in a terrorism response and there is no hole in the ground, it is a hazardous materials problem. Fortunately for HAZMAT teams, these terrorism materials are very similar than some industrial chemicals. The intent behind the release may differ, but handling the chemicals can be easily accomplished by a well-trained and equipped HAZMAT team.

The operations level is considered a basic level, and there are several expanded levels beyond operations. There are specialty areas within the hazardous materials field, all of which are covered by various texts published by Delmar. I encourage further learning through these texts and those listed in the Suggested Readings section at the end of each chapter.

ACKNOWLEDGMENTS

It is a unique experience to write a text such as this, and one of the best parts of the job is acknowledging some of the people who have made an impact in my life or have assisted in some fashion with this text. I can never acknowledge everyone that I would like to, but I will do the best I can. These acknowledgments are in no particular order.

I would like to thank the authors of the Street Stories that begin each chapter and John Eversole for writing the Foreword. All of them have been of great assistance through the years, and I have learned a lot from them. I am very thankful for their contributions to this project.

To all of my friends with whom I teach or travel and to those who have shared their knowledge, I am most appreciative. This text is not my effort alone. I have a great phone book, and I haven't hesitated to use it. When I teach I always learn, and this project also taught me a lot. I have had a unique career and have made some great friends. We have had a lot of fun (my rule # 1) and tried to avoid getting killed (my rule # 2).

Several individuals have made an impact on my career, and I always wanted to write a text to thank them for their assistance. The first is Dave Barker from Johnstown, Pennsylvania, my first chief and instructor. His work ethic, drive, and desire pushed me to try to become not only an instructor, but a great instructor just as he was. The other officers at Richland Township provided a lot of leadership through my years there as well. The officers and firefighters at Monroeville # 4 (Pennsylvania) expanded my learning to a higher level, and it was there that I decided to become a career firefighter. The live-in crew and my EMS partners during that time were great to work with and a lot of fun, too. In the city of Durham, North Carolina, where I first got interested in hazardous materials, Luther Smith, now a battalion chief, was the best officer I have ever worked for—and a good friend. It was with the crew of Engine 3 that I had the experiences of a lifetime. In Baltimore County I am thankful for the opportunity given to me by the Fire Chief John F. O'Neill and Deputy Chief Gary Warren. Chief Warren allowed me to grow and try a lot of these ideas out on our team. I was able to accomplish a lot in my career due to the efforts of Chief Warren, another good friend. All of the shifts at HAZMAT 114 have been of great assistance to me. Thanks to stations 13 and 14 B shift for being in the photographs.

One of the benefits our HAZMAT team has is the assistance of chemist Gene Reynolds, who has been a motivator to learn more about this business, a great relief at many incidents, and a good friend. Buzzy Melton, who is also a chemist and a friend, added his talents to many of our combined training programs. It was during the development of a training program that Buzzy and I developed the risk-based response theory. Buzzy provided the "show me" theory of HAZMAT response and was always willing to give me a hand. In this realm I also need to thank the responders at the Maryland Department of the Environment, an agency that has some of the most knowledgeable and experienced responders in the country.

Many others in the HAZMAT field have earned my respect and become good friends. Two very good friends are Steve Patrick and Tamara Patrick, who both provided some valuable reviews of this text. Steve is responsible for getting me involved in the terrorism side of HAZMAT and for that I am very thankful. The folks at the FBI's HMRU have been of great assistance as well.

My mother and my sons certainly deserve some thanks as they have put up with a lot during this project. My mother planted the idea to pursue my dreams, and my sons have helped me to do so.

Finally, I would like to thank the folks at Delmar for making this project a reality and for believing in me and my various projects. Jeanne Mesick has been valuable to me both as an editor and a friend. She is the ring leader in my inner saga circle of people who are always willing to provide advice no matter how much it hurts. And I do need to thank my saga friends; without them the road show would not have been as fun. This text should be the start of one big saga. I am grateful for all of the assistance that my friends have provided over the years. Without them this would not have been possible. HAZMAT has given me the opportunity to travel all across the United States and be involved with some of the best and the brightest.

I will leave you with some simple rules: There are three things that can't be said when discussing hazardous materials—never, always, and best. Stay flexible, and when told something, always respond, "Show me." Be safe, apply rules #1 and #2 . . . and continue to learn.

REVIEWERS

The author and Delmar gratefully acknowledge the time and effort of the many individuals who reviewed this material and offered their suggestions and expertise in an objective manner. A condensed version of this text appears in *The Firefighter's Handbook: Essentials of Firefighting and Emergency Response.* The review panel at that time pointed out to us that we had the makings of another opportunity to teach those who wanted to get to take their hazardous materials learning even further. Our special thanks to those reviewers:

Fred C. Windisch, Fire Chief, Ponderosa VFD, Houston, Texas, and Volunteer Chief Officer Section—International Association of Fire Chiefs Chairman.

John J. Salka, Jr., Battalion Chief, New York City Fire Department

Michael J. Connors, Assistant Fire Chief, Naperville Fire Department, Naperville, Illinois

David P. Pritchett, Director, Georgia Fire Academy, Forsyth, Georgia, and Training Captain, City of Jackson, Jackson, Georgia

Chris Reynolds, Battalion Chief, Hillsborough County Fire Rescue, Tampa, Florida

Randy Scheerer, Battalion Chief, Newport Beach Fire Department, Newport Beach, California

William Shouldis, Deputy Chief, Philadelphia Fire Department, Philadelphia, Pennsylvania

Robert Bettenhausen, Chief Fire Marshal, Village of Tinley Park, Tinley Park, Illinois

Lee Cooper, Fire Service Specialist, Wisconsin Indianhead Technical College, New Richmond, Wisconsin, and President, Fire Instructors Association of Minnesota, Bloomington, Minnesota

As the material was being further developed to deliver the text you are reading now, we built yet another panel of experts to further enhance the text. Again our special thanks go to:

Chip Osborne, Davidson County Community College, Fire/Rescue Training Department, Lexington, North Carolina

Clyde Lansing, Fresno City Fire Department, Fresno, California

Paul Morley, Atlanta Fire Department, HAZMAT Unit, Atlanta, Georgia

ABOUT THE AUTHOR

 Christopher Hawley is a Fire Specialist with the Baltimore County Fire Department, currently assigned as the Special Operations Coordinator. Prior to this assignment he was the Hazardous Materials Coordinator, a position he held for eight years. He has been a HAZMAT responder for more than fourteen years. Chris has twenty-one years experience in the fire service and has been with the Baltimore County Fire Department for the twelve years, with assignments in fire suppression, fire and rescue training, and hazardous materials. Prior to that, he was a Hazardous Response Specialist with the City of Durham, North Carolina, Fire Department. He has served as a volunteer in Loganville, Monroeville #4, and Richland Township (Johnstown), Pennsylvania, and has served in a variety of ranks, including chief.

Chris has designed innovative programs in hazardous materials and has assisted in the development of other training programs, many for the Maryland Fire and Rescue Institute (MFRI). Chris is an adjunct instructor to the National Fire Academy and has served on development committees. He currently serves as a co-developer for Emergency Response to Terrorism: Tactical Considerations HAZMAT course and as technical advisor and reviewer of other terrorism programs. Chris has presented at numerous local, national, and international conferences. He serves on a variety of pivotal committees and groups at the local, state, and federal levels. He works with local, state, and federal committees and task forces related to hazardous materials, safety, and terrorism.

Chris is the owner of FBN Training, which provides a wide variety of emergency response training, including hazardous materials, confined space, technical rescue, and emergency medical services. FBN Training also offers consulting services to emergency services and private industry. As a subcontractor to Community Research Associates (CRA) he cowrote the FBI Hazardous Materials Operations training program and regularly serves as a course manager for these programs. Through CRA Chris was a technical consultant for the hazardous materials technician level program and serves as one of three course managers for that training program. Through CRA he has served as course manager for a hazardous materials training program for specialized assets of the FBI nationwide. Through FBN Chris is a technical consultant to a number of air monitor manufacturers, including those that manufacture warfare and terrorism agent detectors. Also through FBN, Chris continues to serve as a technical consultant to a Department of Defense contractor; specifically, he was hired to provide information related to a fire department's response to a terrorist incident.

INTRODUCTION TO HAZARDOUS MATERIALS

Hazardous materials response is one of the specialty fields within the emergency services. When the responders are called to a hazardous materials release, the actual handling of chemical releases is usually handled by a well-trained team whose function it is to handle such emergencies. In almost all cases firefighters and other emergency responders are required to assist in the effort as well.

NOTE Many times at hazardous materials incidents the actions taken in the first five minutes determine how the next five hours/days/months will go.

Emergency responders are bombarded with exposures to hazardous materials each and every day in their regular duties. A large number of emergency responders are exposed to blood-borne pathogens each day, sometimes suffering fatal effects from this exposure. By the nature of our jobs we are exposed to numerous toxic and cancer-causing agents, many times in nonemergency situations. When confronted with emergency situations, the exposure and risk increases as well. This text covers some additional hazardous materials risks and offers some fundamental response profiles that will help keep you safe. As the emergency services looks to further its efforts at improving health and safety, the response to chemical accidents is one area where extra precautions are needed.

SAFETY In a fire situation the injuries can be acute, as when a building collapsing during firefighting operations; by comparison, a chemical exposure can also kill responders immediately with a deadly dose or cause serious illness or injury that prolongs death for ten to twenty years.

As the community became more environmentally aware, so did the emergency services. Hazardous materials response is still a very young field, and most communities have only had fire department hazardous materials response teams for sixteen years or less. The Jacksonville, Florida, team is the oldest in the country at 21 years old. This field is constantly changing, and the response to hazardous materials emergencies has undergone three distinct changes in response methodology. Tactics employed to handle incidents ten years ago certainly were appropriate then, but might bring criticism if used today. Technology is rapidly changing as well; air monitoring devices that were used strictly by hazardous materials response teams now are common on fire engines, medic units, bomb squads and tactical police units. On the hazardous materials response team side, technology has also increased to a level that requires a high degree of training and expertise. The response to chemical accidents is a rapidly changing field and brings new experiences every day. At building fires, tank truck fires, clandestine drug labs, shipboard incidents, pressurized cylinder leaks, overturned tank trucks, and attacks of terrorism, hazardous materials response teams face new challenges on every response. This text is to provide you with the practical framework to keep you and your fellow responders alive in situations that involve hazardous materials.

The next few sections will describe in detail the many aspects of hazardous materials and your actions at those incidents.

In this text we cover the objectives as required by the Hazardous Materials Section, NFPA 1001. This section requires that the student receive hazardous materials training at either the awareness or operations level, which is outlined in NFPA 472, *Standards for Professional Competence of Responders to Hazardous Materials Incidents.* This text will identify which objectives are awareness (A) or operations (O) level. In some cases the objective exceeds the operations level, and in most cases would be at the technician level; however, some of the objectives provided here exceed any NFPA levels. These are noted with an O+, which signifies operations-plus and is not an NFPA level. This information is crucial for your survival but may not be covered by an NFPA objective, or it may be an objective that exceeds the operations level.

CHAPTER 1

HAZARDOUS MATERIALS: LAWS, REGULATIONS, AND STANDARDS

OUTLINE

- Objectives
- Introduction
- Laws, Regulations, and Standards
- Emergency Planning

- OSHA HAZWOPER Regulation
- Standards
- Standard of Care

- Additional Laws, Regulations, and Standards
- Summary
- Review Questions

STREET STORY

We arrived on the scene to find a patient lying beside the hot dusty road, baking in the summer sun. He was a male in his mid-thirties, wearing torn blue jeans, old tennis shoes, and no shirt. A local farmer was on the scene. He was driving home when he noticed the man beside the road. It was obvious he was in distress, so the farmer dialed 911 from a cellular phone.

When we got off the fire engine, it was apparent the man was struggling to breathe. He was sunburned, sweaty, and unable to respond to any of our questions. He was shaking slightly, but it did not look like a typical epileptic seizure. We put the man on high-flow oxygen and began to assess his vital signs. It was the size of his pupils that tipped us off. In addition to the pinpoint pupils, the patient was incontinent and producing volumes of frothy sputum. We stripped off his clothes, decontaminated him with water, and alerted the incoming ambulance by radio that we might have organophosphate poisoning. Fortunately, we had recently had some hazardous materials training on pesticide exposures. Had we not been armed with that information, we might not have picked up on the clues so quickly.

The transport ambulance arrived what seemed like hours later, and by that time we were assisting his respirations with a bag valve mask and suctioning his airway. An IV was started, and atropine was administered in large doses to counteract the effects of a suspected pesticide. Unfortunately, we were never able to identify the substance he was exposed to or even know the duration of the exposure.

Sometimes, despite the greatest efforts and best technology available, people do not survive. He fought hard to breathe and even harder to live, but in the end, neither we nor the hospital staff could save his life. He died an hour or so after we first saw him, probably never knowing why he became so sick.

I think about him from time to time, wondering if the call could have gone differently, but I always arrive at the same conclusion: You do your best, learn from any mistakes or situations you find yourself in, and hope you have a better outcome the next time around.

*We could have suffered a secondary exposure because we were not properly protected, but we did not. We could have had an inhalation exposure from his contaminated clothing, but we did not. We could all have ended up as patients, but we did not. It was only luck that brought us through without harm. Would I do it differently next time? Absolutely. Our first responsibility as firefighters is to not make the problem worse when we show up! You are no good to anyone if you become part of the problem or end up being a victim yourself. Use common sense, wear the right **personal protective equipment (PPE)** for the right situation, and, above all, think before you act. Do not count on luck to bring you back to the station.*

—Street story by Rob Schnepp, Firefighter/Paramedic, Alameda County Fire Department, Alameda County, California

OBJECTIVES

After completing this chapter, the reader should be able to explain:

- The local emergency response plan and standard operating guidelines (A)
- The student's role within these documents at the awareness level (A)
- The notification process to request assistance (A)
- The role of the Local Emergency Planning Committee (O+)
- The role of the SARA Title III regulation and emergency response (O+)
- Other regulations related to fire department activities (O+)

INTRODUCTION

To establish the groundwork for this text, we need to define **hazardous materials (HAZMAT)**. The text box included here provides standard definitions used by federal agencies. An excellent definition was provided by Ludwig Benner, Jr., of the National Fire Academy: *"any substance which jumps out of its container when something goes wrong, and hurts and*

DEFINITIONS OF HAZARDOUS MATERIAL

Department of Transportation (DOT) Hazardous Material: Any substance or material in any form or quantity that poses an unreasonable risk to the safety and health and to property when transported in commerce. The DOT also provides more exacting definitions with each hazard class.

Environmental Protection Agency (EPA) Hazardous Material: Any chemical that, if released into the environment, could be potentially harmful to the public's health or welfare.

Occupational Safety and Health Administration (OSHA) Hazardous Chemical: Chemicals that would be a risk to employees if they are exposed to the substances in the workplace.

---

harms the things it touches." As you will learn, almost everything solid, liquid, or gas is hazardous to human health if it is used improperly or escapes its container, as is demonstrated in **Figure 1-2**. The most toxic material is not hazardous as long as it is used correctly and stays in the container in which it was intended to be transported. Another important principle is that even if the material escapes its container, it must have the potential to harm humans, property, or the environment to be considered hazardous. Hazardous materials are packaged to be transported every day and are intended to stay in that package, as shown in **Figure 1-3**. When responders become involved in an incident

Figure 1-2 An example of hazardous material that ignites when it comes in contact with the air after escaping its container. A material that is air reactive is known as pyrophoric.

Figure 1-3 Hazardous materials team members take out a cleaning valve on a tank truck of sodium hydroxide while first responders stand by with a hoseline.

that involves such a package, their increased level of concern is understandable. But as long as the material remains in the package and the responders do not open it, the risk to them and the environment is minimal. The material may be toxic, flammable, corrosive, or reactive, but it is only harmful if it escapes or is forced from the container.

LAWS, REGULATIONS, AND STANDARDS

First responders should have a basic understanding of the legislative history of hazardous materials. The first regulations applying to emergency responders specified how they would respond and how firefighters would be trained for HAZMAT incidents. Many environmental and safety regulations affect how we respond to emergencies. This section will provide an overview of the major federal laws, regulations, and standards. Readers should consult their local environmental and OSHA offices for more information about state and local laws and regulations.

Development Process

To enhance understanding of this section it is important to know the difference between laws, regulations, and standards. **Laws** are legislation passed by Congress; once signed by the president, legislation becomes law. **Regulations,** on the other hand, are developed by government agencies, like the **Occupational Safety and Health Administration (OSHA)** or the **Environmental Protection Agency (EPA).** Regulations have the weight of law but are not passed by Congress nor signed by the president. In some cases laws require the development of regulations and provide a framework for the government agency to follow. **Standards** are developed by a nongovernmental consensus committee, such as the **National Fire Protection Association (NFPA);** after a public review recommendations from the organization become standards. These do not have the weight of law, but could be applied by a regulating agency or in court. Some standards are being applied with the weight of law, typically by OSHA. As citizens and an industry, it is important for emergency responders to participate in the development and review of laws, regulations, and standards. Many states now have adopted NFPA standards into health and safety regulations, which cover not only industry but now can be applied as a regulation to a fire department. The catchall for occupational safety regulators is known as the general duty

clause: which means that all employers have a general duty or an obligation to provide a workplace free from hazards. To apply the general duty clause the enforcement agency demonstrates that the standard is a consensus standard for the fire service and is for the most part universally adopted as a national standard. Failure to follow the standard, which provides objectives that may have prevented the injuries or deaths, allows the enforcement agency to apply the general duty clause. The regulatory agency does not need a standard to apply the general duty clause, but it certainly strengthens its case against the alleged violator.

EMERGENCY PLANNING

The first law that regulated how emergency services organizations respond to emergencies was the **Superfund Amendments and Reauthorization Act,** commonly referred to as **SARA.** This law was passed in 1986 to protect emergency responders and the community with the intent to inform emergency responders and area residents of chemical hazards within their community. The main component of the law is called the **Emergency Planning and Community Right to Know Act (EPCRA).** It is divided into two sections: planning for emergencies and a mechanism to provide chemical storage information to emergency responders.

State and Local Emergency Response Committees

The planning portion established a requirement to provide **State Emergency Response Committees (SERC)** in each state. The SERC is responsible for ensuring the state has the resources necessary to respond safely and effectively to chemical releases. The SERC also provides the framework to implement **Local Emergency Planning Committees,** known as **LEPC.** The LEPC is a group comprised of the representatives of the community, emergency responders, industry, hospitals, media, and other government agencies. Most LEPCs are set up on a county basis, although larger municipalities may establish their own committee. In some states each county has an LEPC, while some LEPCs represent entire states. If the fire department is not involved with the LEPC, then the local emergency management office should be the point of contact for this committee. The LEPC, as shown in **Figure 1-4**, is responsible for the development of a hazardous materials emergency plan and its annual revision.

LOCAL EMERGENCY PLANNING COMMITTEES

The importance of an emergency services or HAZMAT team's involvement in an LEPC cannot be understated. The LEPC meeting is a low-key event to interact with safety and environmental professionals in the community. The LEPC has a legal mandate to fulfill regardless of the HAZMAT team's involvement. The emergency plan could dictate how the HAZMAT team performs its job on a chemical release. The LEPC is required to exercise the emergency plan every year, and this is a great time to showcase the HAZMAT team's talents to the community. Many LEPCs are involved with community environmental and outreach events, further improving community relations. For example, the Baltimore County Department of Environmental Protection conducts a Household Hazardous Waste Day on an occasional basis. The LEPC, bomb squad, and HAZMAT team are involved in assisting with the collection of hazardous waste. In addition to providing a public showcase, it is a great learning opportunity for the emergency responders as a variety of hazardous waste usually shows up. The Baltimore County LEPC has taken a proactive approach to the business community to make sure all businesses are compliant with state and federal regulations. The fire department is not the enforcement authority in these matters but assists in the completion of forms and conducts inspections looking for any problem areas. The fire department works with the businesses to achieve their compliance and avoid their being fined by other agencies. This work is done under the guidance of the LEPC and is well received by the business community. The LEPC also conducts annual seminars prior to filing deadlines for chemical inventory reporting and hosts an annual business seminar on environmental and safety issues.

Figure 1-4 The LEPC should be a cross section of the community: emergency responders, citizens, and industry. The size of the community should dictate the frequency of the meetings, but they should be held several times a year.

Local Emergency Response Plans

The emergency plan is an important component of a successful response. The emergency plan should outline emergency contacts and procedures. Target facilities such as those that store extremely hazardous substances are to be included with specific information and response tactics. A decision tree for evacuation and in-place sheltering is usually included. If the decision is made to evacuate citizens, the plan provides the mechanism to shelter and take care of evacuees. It is crucial that personnel understand this plan and know how to access its resources during an emergency.

The LEPC is also responsible to make sure local resources are adequate to respond to a chemical release in the community and to serve as the focal point for community awareness. Ensuring that local responders receive the proper amount of training and that the emergency plan is evaluated annually by conducting an exercise (drill) is also part of the committee's responsibility. It is important for the emergency services to be an integral player in the LEPC. Decisions regarding your response to chemical incidents may be planned at the LEPC level. Many of the facilities that you respond to will have representatives at these meetings, and it is better to meet these representatives before, rather than during, an emergency. Although each state has different policies, in many locations funding for training and planning is available through an LEPC since most federal HAZMAT grants are generally provided through the LEPC system.

One of the most widely known portions of the local emergency plan is incident levels. Some jurisdictions have outlined incident levels described in **Table 1-1** to indicate the severity of incidents. This is usually outlined in the local emergency plan and duplicated in the HAZMAT team policy.

Chemical Inventory Reporting

The other section of EPCRA, usually referred to as SARA Title III, covers the chemical reporting portion of the act. It requires some facilities to report

TABLE 1-1

	Incident Levels
Level	**Incident scale**
1	Small scale incident that can usually be handled by the first responders but may also require minimal additional resources. Notifications are usually local and may only be internal to the fire department. The highest level of PPE required is firefighters' turnout gear and self-contained breathing apparatus (SCBA), and that may not be necessary. The material is not toxic by skin absorption, although it may have an inhalation hazard. The size of the spill is usually small and will have minimal environmental impact. Incidents of this type include natural gas or propane leaks and small fuel spills.
2	Incidents at this level usually require additional assistance, such as by the HAZMAT team. The incident may require additional notifications to environmental or emergency management agencies on a local or state basis. The amount of material may be larger or more hazardous. The type of PPE will probably be chemical protective clothing or other clothing not carried by the first responders. The first responders may have switched from an active role to a support role. A level 2 incident may require a small evacuation and a larger isolation incident. The incident will also probably hamper the ability of the jurisdiction to respond to other emergencies, depending on available resources. An example incident would be an overturned gasoline tanker or a leaking propane tanker. A leaking drum in the back of a tractor trailer would be classified as a level 2 incident.
3	A level 3 incident requires substantial local resources and the assistance of other local and state agencies. Resources on the federal level may also be required or, at a minimum, notified. The incident will require the evacuation of the affected area and a substantial isolation area. The release is large or the material is extremely toxic. Examples of a level 3 incident include a train derailment with chlorine railcars leaking or a substantial leak from an ammonia tank truck.

chemical information to the state, LEPC, and the local fire department. Failure to report this information on an annual basis can result in fines of $25,000 a day. To qualify as a reporting facility the entity has two methods of reaching a reporting threshold. One criteria to meet the reporting threshold is storing more than 10,000 pounds of a chemical. The 10,000-pound limit is for a single chemical and is not the total of all chemicals on site. A facility that stores drums of ammonia and water is only required to report the ammonia, not the total weight of the drum. It would need to account for all of the ammonia at the facility, no matter the location or container. Retail gas stations are exempt from reporting gasoline or diesel fuel as long as the amount stored is less than 75,000 and 100,000 gallons, respectively. If the gas station stores more than 10,000 pounds of propane or any other chemical, it is required to report. Other facilities that store or use gasoline or diesel are required to report these materials if the quantity exceeds 10,000 pounds (1,300 gallons).

Another method of meeting the threshold of reporting is to store one of 366 chemicals that the EPA considers **extremely hazardous substances (EHSs).** These substances have separate reporting requirements and lower reporting thresholds. Some require reporting at 100 pounds. If these chemicals are released, the facility has a whole host of responsibilities to comply with, including immediate notification to the LEPC, usually by dialing 911 or other listed emergency contact number as determined by the LEPC. Common EHSs are chlorine, sulfur dioxide, anhydrous ammonia, and sulfuric acid. As EHS materials may present an extreme threat to the community, it is important for emergency responders to become familiar with the facilities that use them and to use caution when responding to potential releases involving these materials.

Facilities that are required to report as SARA facilities must submit either a Tier 1 or Tier 2 Chemical Inventory Report. The Tier 2 report, as shown in **Figure 1-5**, is the most common and lists the chemical name, storage amount in a range, storage location and information, and emergency contact information, including 24 hour phone numbers. The facility is also required to submit a list of

Tier Two	Facility Identification ①	Owner/Operator Name	Page ___ of ___ pages

Emergency and Hazardous Chemical Inventory

Specific Information by Chemical

Name _____
Street _____
City _____ State _____ Zip _____

SIC Code _____ Dun & Brad Number _____

For Official Use Only
ID# _____
Date received _____

Name _____ Phone _____
Mail Address _____

Emergency Contact ②
Name _____ Title _____
Phone _____ 24 hr phone _____

Name _____ Title _____
Phone _____ 24 hr Phone _____

Important: Read all instructions before completing form | Reporting Period From January 1 to December 31, 19____ | ☐ Check if information is identical to the information submitted last year.

Chemical Description	④ Physical and Health Hazards	⑤ Inventory	Container Type / Temperature / Pressure ⑥	Storage Codes and Locations (Non-confidential) Storage Locations	Optional
CAS _____ Trade Secret ___ Chem. Name _____ Check all that apply ③ ☐☐☐☐☐☐ Pure Mix Solid Liquid Gas EHS EHS Name _____	Fire ☐ Sudden release of pressure ☐ Reactivity ☐ Immediate (acute) ☐ Delayed (chronic) ☐	☐☐☐ Max Daily Amount (code) ☐☐☐ Avg. Daily Amount (code) ☐☐☐ No. Days on site (days)			☐
CAS _____ Trade Secret ___ Chem. Name _____ Check all that apply ☐☐☐☐☐☐ Pure Mix Solid Liquid Gas EHS EHS Name _____	Fire ☐ Sudden release of pressure ☐ Reactivity ☐ Immediate (acute) ☐ Delayed (chronic) ☐	☐☐☐ Max Daily Amount (code) ☐☐☐ Avg. Daily Amount (code) ☐☐☐ No. Days on site (days)			☐
CAS _____ Trade Secret ___ Chem. Name _____ Check all that apply ☐☐☐☐☐☐ Pure Mix Solid Liquid Gas EHS EHS Name _____	Fire ☐ Sudden release of pressure ☐ Reactivity ☐ Immediate (acute) ☐ Delayed (chronic) ☐	☐☐☐ Max Daily Amount (code) ☐☐☐ Avg. Daily Amount (code) ☐☐☐ No. Days on site (days)			☐

Certification (Read and sign after completing all sections)
I certify under penalty of law that I have personally examined and am familiar with the information in pages one through ___ and that based on my inquiry of those individuals responsible for obtaining information, I believe that the submitted information is true, accurate, and complete.

Name and official title of owner/operator or owner/operator authorized representative | Signature | Date signed

Optional Attachments
☐ I have attached a site plan
☐ I have attached a list of site coordinate abbreviations
☐ I have attached a descriptions of dikes and other safeguard measures

Figure 1-5 This form is an example of what facilities are required to submit to the fire department and LEPC on an annual basis.

chemicals or **Material Safety Data Sheets (MSDSs).** A site plan outlining the storage areas may also be required to be submitted. EHS facilities are also required to submit a copy of their emergency plan. Depending on the state or locality, reporting requirements may be more stringent, although all must meet the minimum listed for the Tier 2 report. It is important to note, however, that the information contained on a Tier 2 form regarding the chemicals located on site is for the prior year. Facilities report from January 1 to March 1, for the year prior.

The basis for this reporting is to inform emergency responders of the hazard of responding to a SARA facility. It also allows emergency responders to make informed decisions as to site entry, tactical decisions, and potential evacuation decisions. As we will learn, MSDSs are a valuable tool in obtaining

chemical information. EPCRA requires their availability to emergency responders. The Hazard Communication Standard (29 CFR 1910.1200) also requires businesses to have MSDSs available for emergency situations.

OSHA HAZWOPER REGULATION

Another section of SARA required OSHA to develop a regulation covering activities that involve hazardous materials. The regulation known as the **Hazardous Waste Operations and Emergency Response** became final on March 6, 1989. It is referred to as **HAZWOPER,** or by its identification number, 29 CFR 1910.120.

INSTRUCTIONS FOR TIER 2 FORM

The Tier 2 form contains a lot of valuable information (**Figure 1-4**). Probably the most valuable is the 24-hour contact information. Some jurisdictions have entered the Tier 2 information into a computer database for easy retrieval. When companies are alerted for a response to a SARA facility, the dispatcher informs them that they are responding to a SARA Title III facility. This alerts the responders to the storage of chemicals. Unfortunately a lot of the information is coded on these forms. The parts of the Tier 2 form include:

1. Full name and address for the reporting facility. A street address is required. The Standard Industrial Classification Code (SIC) identifies the type of facility. A Dun and Bradstreet number is a unique identification number for the business, similar to Social Security numbers for individuals.

2. Emergency contact section. Two contact persons who should be knowledgeable about the chemicals listed or at least have information related to the facility are specified. The phone number is the day phone, and the 24-hour number is for nights and weekends.

3. The chemical description box. The Chemical Abstract Registry Service Number (CAS #) identifies a specific chemical. The chemical name must be clearly provided on the lines below the CAS number. The form indicates whether the material is pure or a mixture and its state (solid, liquid, or gas) and also notes whether or not it is an EHS.

4. Physical and health hazards. All hazard categories that may apply must be checked. The EPA provides the following categories, and the corresponding OSHA guidelines are also provided.

EPA hazard category	OSHA hazard category
Fire Hazard	Flammable
	Combustible
	Pyrophoric
	Oxidizer
Sudden Release of Pressure	Explosive
	Compressed gases
Reactive	Unstable reactive
	Organic peroxide
	Water reactive
Immediate (acute) Health Hazard	Highly toxic
	Toxic
	Irritant
	Sensitizer
	Corrosive
	Other hazardous chemicals having an adverse effect with short-term exposure
Delayed (chronic) Health Hazard	Carcinogens
	Other hazardous chemicals having an adverse effect with long-term exposure

5. Inventory
   The reporting ranges that must be determined are:

	Reporting ranges	
Range code	From	To (in pounds)
01	0	99
02	100	999
03	1,000	9,999
04	10,000	99,999
05	100,000	999,999
06	1,000,000	9,999,999
07	10,000,000	49,999,999
08	50,000,000	99,999,999
09	100,000,000	499,999,999
10	500,000,000	999,999,999
11	1 billion	>1 billion

6. Storage codes and locations. Boxes for codes provide information related to container type, pressure, and temperature. The codes are listed below:

	Storage Types		
Code	Description	Code	Description
A	Aboveground tank	J	Bag
B	Belowground tank	K	Box
C	Tank inside building	L	Cylinder
D	Steel drum	M	Glass bottles or jugs
E	Plastic or nonmetallic drum	N	Plastic bottles or jugs
F	Can	O	Tote bin
G	Carboy	P	Tank wagon
H	Silo	Q	Railcar
I	Fiber drum	R	Other

Pressure conditions

1 = Ambient pressure
2 = Greater than ambient pressure
3 = Less than ambient pressure

Temperature conditions

4 = Ambient temperature
5 = Greater than ambient temperature
6 = Less than ambient temperature
7 = Cryogenic conditions

CASE STUDY

The HAZMAT company was dispatched to a building fire involving chemicals. The first alarm assignment was already on the scene actively extinguishing the fire in the building. The building housed a computer chipboard manufacturing company. The bulk of the fire was in the production area. When the firefighters entered the building to extinguish the fire, they found several open vats of chemicals. Most of the vats had melted down, releasing a substantial amount of product, most of which was laying on the floor of the building. Fire crews had contaminated themselves and would require some type of decontamination. As decontamination is chemical specific, the HAZMAT crews attempted to locate the owner, who would be needed to identify what types of materials were present. The HAZMAT crews were unable to find any chemical storage information specific to that facility. The owner was located but did not have the required information. The materials that were on the floor and in the vats could not be identified through the use of facility paperwork. The crews were held for several hours, wondering what health threat they faced. Had the proper paperwork been provided, there would have been no delay in getting information.

NOTE Federal regulations are published in the Federal Register, issued daily; they are then annually published in a text called the Code of Federal Regulations (CFR). Each federal agency is identified by a two-digit number (29 for OSHA, 40 for EPA, 49 for DOT), and each regulation is given a number to identify it.

Considerable discussion within the fire service when this regulation was established focused on whether or not it applied to all of the fire service. The primary unknown was whether this regulation applied to volunteer firefighters. In some states OSHA only regulates the private sector, and not career firefighters, while in other states only the public sector was subject to this regulation. In many cases volunteer firefighters were determined to be local government employees, since they received worker's compensation benefits or safety equipment from this employer. To end the confusion, the EPA adopted the same regulation and issued it on the EPA's behalf.

ETHICS Since employment is not a concern of the EPA, all persons are covered by HAZWOPER regardless of their employment status.

The EPA's version of HAZWOPER is referenced by the identification number 40 CFR 311. It is interesting to note that OSHA now puts a provision in some of its regulations that specifically mandates the coverage of volunteer employees.

This regulation has had far-reaching effects for the emergency services. It requires certain training and the development of standard operating procedures. It spells out certain requirements in handling chemical releases. It allows only persons trained in the handling of chemical releases to respond to chemical incidents. Only certain training levels allow operation on working in and around chemical releases. Just to be present at a chemical accident requires training, and as responders' activity levels increase so do their needs for further training.

Paragraph q

The majority of this regulation covers the employer's responsibilities at hazardous waste sites. The last section, **Paragraph q,** covers emergency response and applies to the emergency services. It establishes five levels of training, and a requirement for annual refresher training. It requires that an incident command system be used and that an incident commander be present at chemical releases. At larger incidents a safety officer must be appointed. The higher the training levels, the more detailed the requirements. The OSHA two-in-two-out rule originated from the HAZWOPER regulations and was further defined in the Respiratory Protection Regulation (1910.134). The two-in-two-out rule requires that when employees enter a hazardous environment, known to be **immediately dangerous to life or health (IDLH),** they must use the buddy system, which requires at least two persons go into the environment together with a rescue crew standing by. The minimum backup or rescue crew is two, but the rule does not require that, above two, the rescue crew be one for one for those entering the environment. If four persons enter the IDLH atmosphere, a backup crew of two is sufficient; the rule requires that an "adequate" crew be standing by. The thought process is that typically only one person will get trapped and that two will suffice to make a rescue, with the other crew members assisting. The incident commander determines what comprises an adequate rescue crew.

OSHA AND NFPA TRAINING LEVELS

The training levels established by OSHA and the NFPA and their basic responsibilities are as follows:

Awareness—Responders have potential to come across a possible chemical release, identify the potential for a chemical release, call for assistance, and stand by to isolate the area and deny entry to other persons. Persons trained to the awareness level cannot take any action beyond this. This level is intended for police officers, public works employees, and other government employees.

Operations—Persons trained at the operations level can act in defensive fashion to chemical spills. Acting defensively does not extend to entering a hazardous area, but responders can set up dikes, dams, and other containment measures. Training at the operations level allows persons to assist technicians in setting up the various activities required at a chemical incident. Operations training can be expanded to include specialized activities such as decontamination, so that persons trained at this level can assist with this activity. This level is intended for the fire and EMS service.

Incident Commander (IC)—Designed for a person who has received operations level training and training in incident command procedures. The incident commander will be in charge of the incident. To be the incident commander does not require the highest level of chemical response training, but this individual is the senior response official. The IC must rely on the expertise of other responders such as the HAZMAT team, facility officials, or other technical specialists to make strategic and tactical decisions.

Technician—This is the level at which offensive activities can be completed in the hazard area. Other than some specific restrictions outlined in HAZWOPER, there are no general restrictions as to the activities technicians can perform, as long as the activities fit within the scope of their training. Technicians can stop leaks and complete mitigation of the incident. HAZMAT technicians are expected to mitigate or stop the incident from progressing.

Specialist—This level is identified only in HAZWOPER and is intended to denote a person who has received more training than a technician, or someone who specializes in a specific chemical or area of expertise. The training concentrates on chemistry and the identification of unknown material. In some instances specialists operate at an incident and supervise the technicians. The NFPA removed this level in 1992. The NFPA in the 1997 edition of 472 did add several specialty levels, such as tank car specialist. In some cases persons such as heavy equipment operators, chemists, or other persons with specific knowledge may be deemed specialists by the local jurisdiction.

Medical Monitoring

Other sections of this regulation include a requirement for medical monitoring of certain employees. The level of responsibilities dictates whether the employer is mandated to provide employees with an annual physical. The requirements are:

■ Exposure to a chemical above the permissible exposure limit
■ Requirement to wear a respirator or coverage by the OSHA respiratory regulation (29 CFR 1910.134)
■ Injury due to a chemical exposure
■ HAZMAT team membership.

The physician determines the extent of the exam and can establish that exams be given every two years. The only emergency service employees or volunteers required to be given a physical are the members of a HAZMAT team, although OSHA 1910.134 (Respiratory Protection) requires that a medical survey questionnaire be answered by every responder. Answers to the questionnaire determine if a physical is required or not. If a responder is exposed to a hazardous material above the OSHA specified level, the department is required to provide a physical exam. The physician determines the extent of the exam and any other future visit or tests. One common misconception of this section is that HAZWOPER requires that a pre-entry medical exam be provided to those persons who must wear chemical protective clothing. The regulation requires an annual (or biannual, at the physician's discretion) physical to determine fitness for duty; it requires only that EMS be standing by for any entry work. Although it would be a recommended practice to provide pre-entry physicals, there are certain instances that this is not practical. Situations involving victim rescue or rapidly deteriorating situations that require immediate mitigation are two immediate examples. It is always recommended that persons who wear chemical protective clothing (or any protective clothing) be provided a post-entry physical, a concept that should be applied for firefighting operations. To provide some type of recent pre-entry exam many HAZMAT teams have their personnel complete a medical exam as they report to work, which is certainly an option.

STANDARDS

The one group establishing standards that apply most directly to emergency responders is the NFPA. The NFPA establishes a variety of committees that develop a proposed standard, which is then made available for review and comment by the public. After the review process the proposed standard is voted upon by the NFPA committee; if approved, it is the subject of a final public meeting where it is voted on by the group in attendance. Once passed it then is sent to the Standards Council for final action. Only then does it become a standard. As with emergency medical services (EMS), in which an acceptable standard of care applies, typically based upon a national or regional standard, the NFPA establishes a standard of care. Although one cannot be held criminally liable for violating a NFPA standard, emergency responders may be held civilly responsible. One area in which NFPA standards have been applied as having the weight of a regulation is in the HAZMAT arena. OSHA has used this clause to cite employers for violating a NFPA standard.

NFPA 471

Three primary standards apply to hazardous materials response and training, although others cover the storage and use

NFPA 471
Recommended Practice for Responding to Hazardous Materials Incidents

of chemicals. The first standard is NFPA 471, the *Recommended Practice for Responding to Hazardous Materials Incidents*. This standard provides detailed methods and operational procedures beyond the requirements established by the HAZWOPER regulation.

NFPA 472

The second standard is NFPA 472 *Professional Competence of Responders to Hazardous Materials Incidents*. This

NFPA 472
Professional Competence of Responders to Hazardous Materials Incidents

standard lists objectives required to meet the training levels established by the NFPA and OSHA. OSHA

only established some basic criteria for employers to certify their employees as to their ability to respond to HAZMAT incidents. This NFPA document expands the requirements for employers to certify their employees. The levels mirror those of the HAZWOPER regulation, but in 1992 the NFPA removed the specialist category. The 1997 edition added the following competencies: Private Sector Specialist Employee, Hazardous Materials Branch Officer, Hazardous Materials Branch Safety Officer, Tank Car Specialist, Cargo Tank Specialist, and Inter-Modal Specialist. In addition the 1997 edition includes a Tentative Interim Amendment (TIA) for response to terrorism. This TIA provides some additional competencies in each of the basic levels with regard to terrorism. A TIA adds competencies from outside the normal NFPA process. Those competencies will be voted on during the next scheduled revision cycle for that standard. This text incorporates these objectives at the awareness and operations level.

NFPA 473

The third standard is NFPA 473, *Standard for Competencies for EMS Personnel Responding to Hazardous Materials Incidents*. This standard adds

NFPA 473
Standard for Competencies for EMS Personnel Repsonding to Hazardous Materials Incidents

competencies beyond NFPA 472 regarding EMS issues. It provides for EMS Level I and Level II training. The competencies for Level I allow EMS providers to perform patient care in the cold sector and provide a higher level of patient care information. At Level II, EMS providers can perform patient decontamination and work in the warm sector. The level of training is that of a technician with additional competencies in patient care. See Stilp & Bevelacqua, *Emergency Medical Response to Hazardous Material* 1997, for more information on this topic.

STANDARD OF CARE

As emergency responders, we must abide by a **standard of care.** In years past it was deemed acceptable to wash spilled gasoline down a storm drain. Anyone who does that today could face state and federal charges for violating the Clean Water Act. This standard of care is comprised of the laws, regulations, standards, local protocols, and experience that are detailed throughout this text. Violations of this standard of care are evaluated based on three theories: liability, negligence, and gross negligence.

When operating in and around hazardous materials, liability becomes a very real issue. Other than

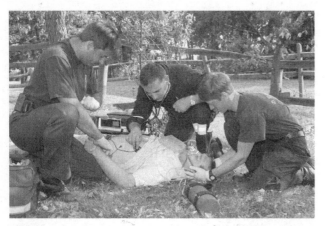

Figure 1-6 Just as EMS responders must follow a standard of care, HAZMAT response has a similar standard.

EMS responses like those shown in **Figure 1-6**, any actions at a chemical release are most likely to put responders at risk as personally liable for any wrongdoing. **Liability** is responsibility for one's actions. Environmental and safety regulations are designed to place blame on the person who made the decision, not on the organization itself, unless the organization has policies that violate these regulations. In the earlier example, the person who made the decision to wash the gasoline down a storm drain could end up in jail and/or pay substantial fines.

Negligence results from not following the standard of care or an accepted practice. **Gross negligence** is willful disregard for the standard of care. In other words a conscious decision is made to not follow the standard of care. To avoid becoming liable, you should receive the training required for your position and act only to your level of training. When you are not sure of your next step, it is best to consult with your local HAZMAT team or other local or state environmental protection agency.

ADDITIONAL LAWS, REGULATIONS, AND STANDARDS

Responders should be aware of the items listed here, as they are commonly encountered or applied in chemical releases.

Hazard Communication

The hazard communication regulation issued by OSHA (29 CFR 1910.1200) requires that employers make available Material Safety Data Sheets (MSDS) for all chemicals located at the facility at quantities above "household quantities." The term *household quantities* provides that small amounts such as those that would be purchased for home use would not require MSDS. An example would be a gallon of household ammonia for use in an office. If a case of household ammonia is brought in, it would require an MSDS as this is not a normal household quantity. The employer also must provide a list of these MSDSs and specify when the chemicals arrived on site. The employer must provide training on these MSDSs and the hazard communication program. One of the important components of this program is the requirement for the facility to make these sheets available for emergency responders. Emergency responders are also responsible for following this regulation.

RESPONDER FACT Emergency response organizations in many states are required to abide by the federal or state environmental and safety regulations. Emergency response organizations are no different than local industry. Even if an emergency response organization is exempt from following environmental or safety regulations, department administrators should consider the public perception of not following the regulations. In many cases the emergency response organization is a regulating entity, which may issue and enforce regulations that local businesses are required to follow. The department's public image is improved if it voluntarily follows these regulations. An added benefit is that responders are working in a safer workplace and protecting the environment.

Superfund Act

The Comprehensive Environmental Response, Compensation, and Liability Act (CERCLA) is commonly referred to as the Superfund law. This law was established for the cleanup of toxic waste sites around the country. It laid the groundwork for regulating the emergency services with regard to response to chemical emergencies. When responding to a Superfund site, some additional concerns and requirements must be noted. The emergency responders should have a copy of the site's existing emergency response plan. The site should have its access limited and hazard zones established; depending on the nature of the work involved, it may have a decontamination corridor. Prior to work beginning on a Superfund site, the local responder sshould receive notification and should meet with the site supervisor to learn the hazards of the site and any protective measures that may be in place. As Superfund sites vary from old landfills to mercury dumping sites, it would be impossible here to provide a breakdown of the potential site hazards and precautions that the site may hold. Prior to becoming a Superfund site, the area may be listed as a National Priority List (NPL) site. Contact your state environmental agency or federal regional EPA office to learn of any Superfund or NPL sites in your area.

Clean Air Act

The Clean Air Act Amendments (CAAA), which were passed in 1990, have some language that applies to the fire service. The CAAA requires facilities that meet thresholds of reporting to file additional planning documents, and the LEPC and local fire service must be involved in training and exercises. In June 1999 the facilities were required to submit emergency plans, many of which must be coordinated with the local fire department. Other requirements, such as exercises and worst-case scenarios, may involve the local responders.

Respiratory Protection

OSHA's respiratory protection regulation (29 CFR 1910.134) also affects the emergency services by adding requirements that had previously been exempt. The biggest changes are the inclusion of the two-in-two-out rule, which was in place for chemical incidents but now also includes structural fire situations. OSHA has now classified a fire as an IDLH situation and requires the incident commander to take certain precautions to ensure firefighters' safety. The emergency response organization is now required to fit test all responders and provide a Licensed Health Care Provider's (LHCP's) approval to wear respiratory protection by the use of a medical survey or a physical exam. Record keeping has also changed; specific records must be kept by the response organization regarding the employees and the respiratory protection program. Records include the listing of personnel training, daily equipment checks, periodic maintenance, routine service, review of the respiratory protection program, and other items.

Firefighter Safety

NFPA 1500 *Standard on Fire Department Occupational Safety and Health Program* is sometimes referred to when discussing HAZMAT issues.

NFPA 1500
Standard on Fire Department Occupational Safety and Health Program

This standard is a broad-based program for providing a safe workplace for firefighters. It has been applied by OSHA using the general duty clause, specifically regarding two-in-two-out.

NFPA Chemical Protective Clothing

NFPA 1991–1994 are standards for chemical protective clothing, including vapor protective suits, splash garments, support garments, and terrorism response garments. They establish design and use requirements. More information on this topic is found in Chapter 4.

SUMMARY

The maze of laws, regulations, and standards can be confusing and sometimes overwhelming. Most laws and regulations are not easy to read, and it can take years to comprehend every detail of their intent. They are subject to interpretation and change frequently, and emergency responders must keep abreast of those that affect their everyday jobs. In larger departments staff may be assigned to monitor health, safety, and environmental laws and regulations. In smaller departments this may not be possible, but as an industry, the responders are the same as the chemical plant or gas station in a community. The responders must follow the same health, safety, and environmental laws and regulations. For your health and safety, it will benefit you to keep abreast of these issues, and it will help make your community a safer place to live and work.

KEY TERMS

Awareness   The basic level of training for emergency response to a chemical accident, the ability to recognize a hazardous situation and call for assistance.

Emergency Planning and Community Right to Know (EPCRA)   The portion of SARA that specifically outlines how industries report chemical inventory to the community.

Environmental Protection Agency (EPA) Federal agency ensuring protection of the environment and the nation citizens.

Extremely Hazardous Substances (EHSs) List of 366 substances that the EPA has determined present an extreme risk to the community if released.

First responders A group designated by the community as those who may be the first to arrive at a chemical incident. This group is usually comprised of police officers, EMS providers, and firefighters.

Gross negligence Disregarding training and acting without regard for others.

Hazardous materials (HAZMAT) Those substances that have the ability to harm humans, property, or the environment.

Hazardous Waste Operations and Emergency Response (HAZWOPER) The OSHA regulation that covers safety and health issues at hazardous waste sites and response to chemical incidents.

Incident Commander HAZMAT training level that encompasses the operations level with the addition of incident command training. Intended to be the person who may command a chemical incident.

Laws Legislation approved by Congress and signed by the president.

Liability Being held legally responsible for your actions.

Local Emergency Planning Committee (LEPC) Group comprised of members of the community, industry, and emergency responders to plan for a chemical incident and to ensure that local resources are adequate to handle an incident.

Material Safety Data Sheets (MSDS) Information sheets for employees that provide specific information about a chemical, with attention to health effects, handling, and emergency procedures.

National Fire Protection Association (NFPA) A group that issues fire and safety standards for industry and emergency responders.

Negligence Acting in an irresponsible manner, or different from the way one was taught to act, differing from the standard of care.

Occupational Safety and Health Agency (OSHA) Federal agency responsible for enacting safety regulations protecting the nation's workers.

Operations Mid level of HAZMAT training above awareness; provides the foundation that allows for the responder to perform defensive activities at a chemical incident.

Paragraph q Section within HAZWOPER that outlines the regulations governing emergency response to chemical incidents.

Personal protective equipment (PPE) Clothing and equipment designed to protect people responding to hazardous materials incidents; PPE ranges from earplugs to a fully enclosed chemical suit.

Regulations Rules developed and issued by a governmental agency, which have the weight of law.

Specialist HAZMAT training level that provides for a specific type of training, such as railcar specialist; a responder with a higher level of training than a technician.

Standard of Care Combination of training, experience, standards, and regulations that guide a responder to a specific action.

Standards Usually developed by consensus groups establishing a recommended practice or standard to follow.

State Emergency Response Committee (SERC) Group that ensures the state has adequate training and resources to respond to a chemical incident.

Superfund Amendments and Reauthorization Act (SARA) Law that regulates a number of environmental issues, predominately for the chemical inventory reporting by industry to the local community.

Technician High level of HAZMAT training that allows specific offensive activities to take place to stop or handle a chemical incident.

REVIEW QUESTIONS

1. How does the level of response to hazardous materials differ from a structural fire response?

2. How can hazardous materials be easily defined?

3. Who is to receive the chemical reporting required under SARA Title III?

4. Who is responsible to ensure the proper response to a hazardous materials emergency?

5. Which regulation covers all emergency responders who respond to chemical accidents?

6. What is the difference between negligence and gross negligence?

7. Which regulation requires employers to maintain Material Safety Data Sheets?

Suggested Readings

Noll, Gregory, Michael Hildebrand, and James Yvorra, *Hazardous Materials Managing the Incident,* Fire Protection Publications, Oklahoma University, 1995.

Schnepp, Rob, and Paul Gantt, *Hazardous Materials: Regulations, Response, and Site Operations,* Delmar, a division of Thomson Learning, Albany, NY, 1999.

Smeby, L. Charles (editor), *Hazardous Materials Response Handbook,* 3rd ed., National Fire Protection Association, 1997.

Stilp, Richard, and Armando Bevelacqua, *Emergency Medical Response to Hazardous Materials Incidents.* Delmar, a division of Thomson Learning, Albany, NY, 1997.

CHAPTER 2

RECOGNITION AND IDENTIFICATION

OUTLINE

- Objectives
- Introduction
- Location and Occupancy
- Placards, Labels, and Markings
- Other Identification Systems
- Containers
- Senses
- Chemical and Physical Properties
- Summary
- Key Terms
- Review Questions

STREET STORY

While the news media tends to focus on the large and spectacular, the reality is that every day the fire service responds to literally hundreds of incidents involving hazardous materials. These incidents usually do not get much attention because they are small in scope and are handled with a minimal amount of fire department resources. Common examples include natural gas leaks and flammable and combustible liquid spills.

We have an old saying in the HAZMAT community that the initial ten minutes of an incident will dictate the tone for the first hour. Clearly, the actions—or inactions—of the first responders will set the tone for an incident. I think back to two incidents in my career that reinforce this point.

The first incident involved an engine company sent to investigate a call regarding an unknown liquid spilled along a highway. Upon arrival, the engine company officer found a very viscous red liquid spilled along a long distance of road, apparently from the back of a tractor trailer. Unsure of the identity and the potential hazards of the liquid, the officer requested that a HAZMAT unit respond to the scene and provide assistance. Using their monitoring and detection equipment, responders were eventually able to determine the liquid posed no threat to the community. In fact the unknown liquid was eventually identified as strawberry syrup! Although the officer took some ribbing from his peers, he clearly made some proper decisions to ensure the safety of both his personnel and the community.

In the second incident, a police officer was called to provide assistance to a public works road crew that had discovered what appeared to be a 5-pound portable extinguisher wrapped in duct tape with a fuse on top. Although he believed it to be a hoax, the officer still requested the response of the fire department bomb squad. After conducting a thorough risk assessment and requesting the additional assistance of a HAZMAT response team, the bomb squad determined that the perceived hoax was in fact an actual explosive device. When the device was disrupted (i.e., blown up) by the bomb squad, it made a lasting impression on the police officer!

In both of these incidents, recognition and identification by the first responder set the tone for the incident. If you do not know what to do, isolate the area, deny entry, and call for help.

—Street Story by Greg Noll, Emergency Planning and Response Consultant, Hildebrand, Noll Associates, Lancaster, Pennsylvania

OBJECTIVES

After reading this section the student should be able to identify and explain:

■ The nine hazard classes as defined by DOT (A)

■ The hazards associated with each hazard class (A)

■ Standard occupancies where hazardous materials may be used or stored (A)

■ Standard container shapes and sizes and common products (A)

■ Both facility and transportation markings and warning signs (A)

■ The standard transportation types for highway and rail (O)

■ The use of the NFPA 704 system (A)

■ The use of transportation containers in identifying possible contents (O)

■ The location of emergency shutoffs on highway containers (O)

■ The importance of understanding chemical and physical properties of hazardous materials (O)

INTRODUCTION

This chapter is of primary importance to the emergency responder.

SAFETY By honing your skills in recognition and identification (R & I), you can improve your ability to stay alive.

The inability to recognize the potential for chemicals and the inability to identify a chemical hazard can place yourself and others in severe danger. Public saftey is inherently dangerous, but the response to hazardous materials release places an additional risk. Not only can there be immediate effects from some materials, but multiple exposures can have long-reaching effects. In rare cases, fires have killed hundreds of people. Equally rare but even more deadly, releases of extremely hazardous substances have resulted in the deaths of thousands of people in one incident. In 1984 the release of methyl isocyanate in Bhopal, India, killed more than 2,000 people and injured thousands more. This tragedy was the basis for the Emergency Planning and Community Right to Know Act (EPCRA). Several facilities in the United States use this material.

SAFETY There are four basic clues to recognition and identification: location and occupancy, placards labels and markings, container types, and the senses.

Just a suspicion in any of these areas should be enough to place a first responder on guard for the possibilities of a chemical release and associated hazards, as shown in **Figure 2-2.**

LOCATION AND OCCUPANCY

CAUTION The size of the community does not determine the potential for hazardous materials. Every community has hazardous materials.

Most communities have a gas station or a hardware store. The average home has a large amount of hazardous materials, which can cause enormous problems during a response. In rural communities, farms present unique risks due to the storage of pesticides and fertilizers, as shown in **Figure 2-3.** All of these locations and occupancies provide the potential for the storage of hazardous

Figure 2-2 Basic clues to recognition and identification are location and occupancy, placards, labels, markings, container types, and senses. Here, the location is an industrial complex, the containers are visible upright storage tanks, which are labeled with their contents, and all of this is plainly visible.

Figure 2-3 Agricultural supply stores have a large quantity of hazardous materials, including pesticides, herbicides, and fertilizers. In many cases they also stock propane and other fuel.

materials. In general the more industrialized a community, the more hazardous materials the community will contain. Communities adjacent to industrialized areas or along a major transportation corridor (interstate highway, rail, water) will also have the same hazards as these materials will be traveling in the community, as shown in **Figure 2-4.** Buildings that typically store hazardous materials include: hardware stores, hospitals, auto part supply stores, dry cleaners, manufacturing facilities, print shops, medical offices, photo labs, agricultural supply stores, semiconductor manufacturing facilities, electronics manufacturing, light to heavy industrial plants, marine terminals, rail yards, airport terminals and fueling areas, pool chemical stores, paint stores, hotels, swimming pools, food manufacturing facilities, and many others.

PLACARDS, LABELS, AND MARKINGS

In this section we will examine the first concrete evidence of the presence of hazardous materials. A number of systems are used to mark hazardous materials containers, buildings, and transportation vehicles. The systems are part or combinations of laws, regulations, and standards. As an example the **Building Officials Conference Association (BOCA)** code, which is adopted as a regulation in some local communities, requires the use of the NFPA 704 marking system (a standard) for certain occupancies, as shown in **Figure 2-5.**

Placards

This is the most commonly seen system for identifying the location of hazardous materials. The **Department of Transportation (DOT)** regulates the movement of hazardous materials (dangerous goods in Canada) by air, rail, water, roadway, and pipeline, per 49 CFR 170-180. After meeting certain guidelines a shipper must placard a vehicle to warn of the storage of chemicals on the mode of transportation (see **Figure 2-6**).

NOTE The quantity of hazardous materials that must be carried to require placarding is 1,001 pounds, unless it is one of six classes of materials that require placarding at any amount.

Table 2-1 provides a further explanation of the placarding system.

Figure 2-4 If a road goes through your community, the potential for a hazardous materials incident exists. One of the most common chemical releases is a gasoline spill.

Figure 2-5 The NFPA has developed its 704 system of identifying potential chemical hazards in a building.

Figure 2-6 The DOT requires some shippers of hazardous materials to provide warning placards to warn responders of chemicals that may be on the truck. This truck contains several different materials. The corrosive placard with the number "3096" indicates that part of the shipment is in bulk containers.

NOTE The DOT has established a system of nine hazard classes and has more than twenty-seven placards to identify a shipment.

The idea behind these hazard classes is to identify a shipment per a general grouping and provide some basic information regarding the potential hazards. The placards are signs that are 10¾ inches by

Figure 2-7 Corrosive placard.

10¾ inches; they are to be placed on four sides of the vehicle, like the one shown in **Figure 2-7.** Labels are 3.9 inches by 3.9 inches and are affixed near the shipping name on the container. The labels are smaller versions of the placard and are designed to provide warnings on the package's contents.

The system is designed so that materials designated by the DOT that present a capacity to harm the environment, humans, and animals are easily identified. Materials must present an unreasonable risk to the safety, health, or property upon contact. The DOT has established two tables, with different guidelines for both tables. Materials listed in Table 1 are those most hazardous and require the use of placards no matter the quantity of material being carried. Table 2 materials are those that require placarding at 1,001 pounds. The criteria establishing this 1,001 pound rule is not clearly defined but responders should be aware that a spill of 999 pounds of a material can be as hazardous as 1,001 pounds. The shipper uses the hazardous materials table (49 CFR 107.101) to determine which labels and placards are required. The table may also list a packing group for the material, which indicates the danger associated with the material being transported.

■ Packaging Group I—Indicates greatest danger
■ Packaging Group II—Medium danger
■ Packaging Group III—Minor danger

TABLE 2-1(A)

Placards—Requires Placarding at Any Amount	
Hazard Class or Division	**Placard Type**
1.1	Explosives 1.1
1.2	Explosives 1.2
1.3	Explosives 1.3
2.3	Poison gas
4.3	Dangerous when wet
6.1 (assigned to Packing Group (PG) I, inhalation hazard only) Zone A or B	Poison inhalation hazard
7 (radioactive label III only)	Radioactive
5.2 Organic Peroxide, Type B liquid or solid, temperature controlled	Organic peroxide

TABLE 2-1(B)

Requires Placarding at 1,001 Pounds	
Class or Division	**Placard Type**
1.4	Explosives 1.4
1.5	Explosives 1.5
1.6	Explosives 1.6
2.1	Flammable gas
2.2	Nonflammable gas
3	Flammable
Combustible liquid	**Combustible**
4.1	Flammable solid
4.2	Spontaneously combustible
5.1	Oxidizer
5.2 (other than Type B liquid or solid, temperature controlled)	Organic peroxide
6.1 (Packing Group (PG) I or II, other than Zone A or B)	Poison
6.1 (PG III)	Keep away from food
6.2	None
8	Corrosive
9	Class 9
ORM-D	None

TABLE 2-2

Packing Groups: Flammability Class 3		
Packing Group (PG)	**Flash Point**	**Boiling Point**
I		≤ 95°F
II	< 73°F	> 95°F
III	≥ 73°F and ≤ 141°F	> 35°F

Packaging groups are only assigned to classes 1, 3 through 6, 8 and 9 as shown in **Table 2-2, Table 2-3, Table 2-4,** and **Table 2-5.** These are determined based on flash points, boiling points, and toxicity. See **Table 2-2** for more information. At this point in the text these items may be confusing; these terms are further explained toward the end of this chapter, but to assist with this section, they can be defined as:

■ Flash point—the temperature of a liquid that, if an ignition source is present, provides sufficient vapors to start a fire.

■ Boiling point—the temperature of a liquid that is changing from the liquid state to the gaseous state.

■ Toxicity—degree of harmfulness, as provided through exposure limits. These exposures are provided in lethal concentrations (LC) and lethal doses (LD), which measure the amount of material needed to produce toxic effects.

Placards provide four useful methods of communicating the hazard class they represent. They have distinct colors and include a picture at the top of the square depicting a representation of the hazard.

TABLE 2-3

Packing Groups: Poisonous Gas Division 2.3	
Hazard Zone	**Inhalation Toxicity—Lethal Concentration (LC_{50})***
Hazard Zone A	LC_{50} less than or equal to 200 ppm
Hazard Zone B	LC_{50} greater than 200 ppm and less than or equal to 1,000 ppm
Hazard Zone C	LC_{50} greater than 1,000 ppm and less than or equal to 3,000 ppm
Hazard Zone D	LC_{50} greater than 3,000 ppm and less than or equal to 5,000 ppm

*LC_{50} is the lethal concentration to 50% of an exposed population (gases)

TABLE 2-4

Packing Group Division 6.1			
For packing materials that are toxic by a route other than inhalation, the following are used:			
Packing group	**Oral toxicity— Lethal Dose (LD_{50})* (mg/kg)**	**Dermal toxicity— Lethal Dose (LD_{50}) (mg/kg)**	**Inhalation by dusts and mists (LC_{50}) (mg/L)**
I	≤ 5	≤ 40	≤ 0.5
II	$> 5, \leq 50$	$> 40, \leq 200$	$> 0.5, \leq 2$
III	Solids: $> 50, \leq 200$; liquids: $> 50, \leq 500$	$> 200, \leq 1000$	$> 2, \leq 10$

*(LD_{50}) Lethal Dose to 50% of the exposed population (solids and liquids)

TABLE 2-5

Packing Group Division 6.1	
Poisonous by inhalation	
Packing group	**Vapor concentration and toxicity**
I (Hazard Zone A)	$V \geq 500\ LC_{50}$ and $LC_{50}\ 200\ mL/M^3$

MEASURING TOXICITY

The DOT uses several terms to provide a measure of toxicity. The DOT establishes ranges for the shipment of toxic materials. The determination of how toxic a material is can be based upon exposure limits established for many chemicals. LC_{50} and LD_{50} are terms used to describe a concentration or amount of a material that causes lethal effects to 50 percent of an exposed population. LC is lethal concentration, used to measure toxicity of gases, and LD is a lethal dose of solids and liquids. To determine a LC_{50} or LD_{50}, researchers expose a population of test animals to a given amount of a substance. At the level that 50 percent of the test population dies, this establishes the LC_{50} or LD_{50}. When discussing toxic-ity, two common units of measure for the amount of material required to cause toxic effects are parts per million (ppm) and milligrams per meter cubed (mg/M^3). The use of ppm usually is reference to a gas and represents a part of a million pieces, as one marble is one part per million of a million marbles. The use of mg/M^3 usually refers to the amount of a solid or liquid material. It is a comparison of the amount of chemical in milligrams (mg) as compared to a cubic meter of air, as that amount of material could cause toxic effects within that cubic meter. The use of mg/L is the reference to the amount (mg) that is within the space of a liter. More information related to toxicity is provided in Chapter 4.

Placards state the hazard class across their middle and display the hazard class and division number at the bottom.

As an example the placard as shown in **Figure 2-7** is black and white, which represents the corrosive class. The top of the placard has a picture of a hand and a steel bar being eaten away by the corrosive material being poured on it. The middle of the placard states, "corrosive," and at the bottom is the class number 8. The DOT also requires the addition of a four-digit number known as the United Nations/North America (UN/NA) identification number either on a placard or on an adjacent orange strip. This identifies a bulk shipment of over 119 gallons and specifies what the material is. A tank truck carrying gasoline, which would be considered a bulk shipment, would display a flammable placard with the number 1203 either in the middle of the placard or on an orange strip adjacent to the placard as shown in **Figure 2-8.** A tank truck placarded for gasoline could also haul diesel fuel or fuel oil under the same placard without having to change placards. The DOT assumes that if the truck is placarded for the worse hazard, responders will be adequately warned. This panel with the DOT ID number, as shown in **Figure 2-8,** provides an additional bit of information to the responder, as before the only information would be from the placard indicating a flammable liquid was on board.

The nine hazard classes and subdivisions[1] are:

■ **Class 1 Explosives** (placards and labels shown in **Figure 2-9**)

■ Division 1.1—mass explosion hazard, such as black powder, dynamite, ammonium perchlorate, detonators for blasting, and RDX explosives.

Figure 2-9 "Explosive" placards and labels.

■ Division 1.2—projectile hazard, such as aerial flares, detonating cord, detonators for ammunition, and power device cartridges.

■ Division 1.3—fire hazard or minor blast hazard; examples include liquid fuel rocket motors and propellant explosives.

■ Division 1.4—minor explosion hazard, which includes line throwing rockets, practice ammunition, detonation cord, and signal cartridges.

■ Division 1.5—very insensitive explosives that present mass explosion potential but during normal shipping would not present a risk. Ammonium nitrate and fuel oil (ANFO) mixtures are an example of this division.

■ Division 1.6—very insensitive explosives with no mass explosion potential. Materials that present an unlikely chance of ignition are part of this grouping

Incidents involving explosives can be very dangerous, especially when involved in fire. Making a tactical decision to attack a fire involving explosives can endanger the responders, especially if the fire has reached the cargo area of the vehicle. The

Figure 2-8 There are three ways to signify the four-digit ID number for bulk shipments. The most common is the four-digit DOT identification number in the middle of the placard. The four-digit number in the middle of the placard or on an adjacent panel indicates that the shipment is in bulk quantity.

CASE STUDY

On November 29, 1988, an engine company from the Kansas City, Missouri, Fire Department was dispatched to a reported pickup truck fire. While en route to the incident the engine company was told to use caution as there were reports of explosives involved. When firefighters arrived, they found two separate fires, one in the pickup truck and the other in a trailer. The first engine began to extinguish the fire in the pickup truck and requested a second engine for assistance and the district battalion chief. They attempted to contact the second engine to warn them of the explosives on fire on top of the hill. The second engine arrived and began to attack the fire on top of the hill. That crew requested the assistance of the first engine and also requested a squad for water. From the radio communications with the battalion chief, the crews thought that the explosives had already detonated.

The battalion chief was a quarter of a mile away where he had stopped to talk with the security guards when the explosives detonated. The explosion moved the chief's car 50 feet and blew in the windshield. The blast was heard for 60 miles and damaged homes within 15 miles. The chief requested additional assistance and staged the responding companies. There was a report of more explosives that had not yet detonated on the hill. Luckily, the chief did not let any other responders into the scene as shortly after pulling back, a second explosion went off, reportedly larger than the first one. The next morning a team of investigators went to the site to begin the investigation. They discovered that all six firefighters were killed in the blast, and the explosion was very devastating. Only one engine was recognizable; the other was reduced to the frame rail. The first blast was from 17,000 pounds of ammonium nitrate, fuel oil, and aluminum mixture. It also had 3,500 pounds of ammonium nitrate and fuel oil, commonly referred to as ANFO. The second explosion involved 30,000 pounds of the mixture. The use of ANFO is common throughout the United States for blasting purposes. Ammonium nitrate is commonly used as a fertilizer in the agricultural business and is used on residential lawns. The World Trade Center bombing involved ANFO as the main blast component, while the Oklahoma City bombers used ammonium nitrate and nitromethane (racing fuel). When used improperly the results can be devastating.

recommendations in the **Emergency Response Guidebook (ERG)** should be followed, paying particular attention to the isolation and evacuation distances. When explosives are involved in traffic accidents not involving fire, the actual threat is minimized depending on the circumstances. As long as the explosives were transported legally and as they were intended to be transported, the responders should face little danger. Some cities require an escort and transportation of explosives at nonpeak hours. Spilled explosives materials may present a health hazard if inhaled or absorbed through the skin.

■ **Class 2 Gases** (placard and label shown in **Figure 2-10**)

  ■ Division 2.1—flammable gases which are ignitable at 14.7 psi in a mixture of 13 percent or less in air, or a **flammable range** with air of at least 12 percent regardless of the **lower explosive limit.** Propane and isobutylene are examples of this division.

  ■ Division 2.2—nonflammable, nonpoisonous, compressed gas, including liquefied gas, pressurized **cryogenic** gas, and compressed gas in solution. Anhydrous ammonia, carbon dioxide, liquid argon, and nitrogen are examples.

SAFETY Ammonia is listed by the DOT as being nonflammable, but in reality it is flammable. DOT has criteria to meet the flammability standard, and ammonia does not fit into that criteria. Responders have been killed in fires involving ammonia releases. Ammonia should be considered by responders as being flammable, especially when confined in a building.

  ■ Division 2.3—poisonous gases that are known to be toxic to humans and would pose a threat during transportation. Chlorine and liquid cyanogen are common examples of this division. Gases in this division are also assigned a letter code identifying their toxicity levels. You will learn more about these levels in the section on toxicology. The hazard zones associated with this division are:

    ■ Hazard zone A—LC_{50} less than or equal to 200 ppm

Figure 2-10 "Gas" placards and labels.

Figure 2-11 "Flammable" and "combustible" plac-ards and labels.

100°F can be shipped as a combustible liquid. Examples include diesel fuel, kerosene, and various oils.

CAUTION The fire service usually refers to a flammable liquid as one that has a flash point of 100°F or lower, not 141°F as the DOT classi-fies. Any liquid with a flashpoint of greater than 100°F is normally considered a com-bustible liquid, except in the eyes of the DOT.

- Hazard zone B—LC$_{50}$ greater than 200 ppm and less than or equal to 1,000 ppm
- Hazard zone C—LC$_{50}$ greater than 1,000 ppm and less than or equal to 3,000 ppm
- Hazard zone D—LC$_{50}$ greater than 3,000 ppm and less than or equal to 5,000 ppm

The hazard zones are a quick way to determine how toxic a material is, as hazard zone A is more toxic than B, and so forth. The shipping papers will iden-tify these by the addition of Poison Inhalation Haz-ard (PIH) zone A, B, C, or D.

- **Class 3 Flammable liquids** (placard and label shown in **Figure 2-11**) are those liquids that have a flash point of less than 141°F. Gasoline, acetone, and methyl alcohol are examples.
- Combustible liquids are those with flash points above 100°F and below 200°F. The DOT allows that liquids with a flash point of

- **Class 4 Flammable solids** (placard and labels shown in **Figure 2-12**) also include sponta-neously combustible materials and dangerous-when-wet materials.
- Division 4.1 includes wetted explosives, self-reactive materials, and readily combustible solids. Examples include magnesium ribbons, picric acid, and explosives wetted with water, alcohol, or plasticized.
- Division 4.2 are comprised of spontaneously combustible materials, including pyrophoric materials or self-heating materials. An exam-ple is zirconium powder.
- Division 4.3 dangerous-when-wet material are those that when in contact with water can ignite or give off flammable or toxic gas. Calcium carbide when mixed with water make acetylene gas, which is very flamma-ble and unstable in this form. Sodium is

Figure 2-12 Class 4 placards and labels.

Figure 2-13 Class 5 placards and labels.

another example that can, when wet, ignite explosively. Lithium and magnesium are not as explosive as sodium but will react with water. Magnesium if on fire will react violently if water is used in an attempt to extinguish the fire.

▪ **Class 5 Oxidizers and organic peroxides** (placard and labels shown in **Figure 2-13**)

▪ Division 5.1 is the class assigned to materials that can produce oxygen, which in turn increases the chance of fire and during fires makes the fire burn more intensely. Ammonium nitrate and calcium hypochlorite are examples.

▪ Division 5.2 are the organic peroxides that can explode or **polymerize,** which if con-

tained is an explosive reaction. These are further subdivided into seven types.

Type A—can explode upon packaging; these are forbidden by the DOT, which means they cannot be transported and must be produced on site.

Type B—can thermally explode; considered a very slow explosion.

Type C—neither detonates nor **deflagrates** rapidly, and will not thermally explode.

Type D—only detonates partially or deflagrates slowly; has a medium or no effect when heated and confined.

Type E—shows low or no effect when heated and confined.

Type F—shows low or no effect when heated and confined, and has low or no explosive power.

Type G—is thermally stable and is desensitized.

POLYMERIZATION

A chemical that polymerizes is one that presents an extreme danger to firefighters. Polymerization is a runaway chain reaction that once started cannot be stopped. If the material is contained, it will rupture the container in a violent manner, sending pieces of the container for a considerable distance. Polymerization can be best described as a chain. It starts off as one link, and once the reaction is started, many other links develop and attach themselves to the other links. They quickly duplicate themselves, eventually rupturing the container. The reaction can result in a fire or explosion,

as many chemicals that can polymerize are also flammable liquids. A container of spray foam insulation that is used for sealing spaces in walls around doors and windows is a good example. When the foam hits the air it reacts and rapidly expands to a much greater volume. This reaction is uncontrollable, and will continue to expand for quite some time until the polymerization is completed. A small container that holds a few ounces does not present much risk, but a railcar full of a chemical that may polymerize presents an extreme risk to the community.

Figure 2-14 Class 6 placards and labels.

Figure 2-15 Class 7 placards and labels.

Figure 2-16 Class 8 placards and labels.

■ **Class 6 Poisonous materials** (placards and labels shown in **Figure 2-14**) are divided into two groups: poisonous materials other than gases and infectious substances.

■ Division 6.1—materials that are so toxic to humans they would present a risk during transportation. Examples include arsenic and aniline.

■ Division 6.2—microorganisms or their toxins that can cause disease to humans or animals. Anthrax, rabies, tetanus, and botulism are examples.

Hazard zones associated with Class 6 materials are:

■ Hazard zone A—LC_{50} less than or equal to 200 ppm

■ Hazard zone B—LC_{50} greater than 200 ppm and less than or equal to 1,000 ppm

■ **Class 7 Radioactive materials** are those determined to have radioactive activity at certain levels. Although there is only one placard, the labels shown in **Figure 2-15** are further subdivided into Radioactive I, II, and III, with the level III posing the greatest hazard. The I, II, or III describes the allowable amount of radiation that may be detected outside the shipping container.

■ **Class 8 Corrosives** (placard and label shown in **Figure 2-16**) include both acids and bases. The DOT describes corrosives as materials capable of causing visible destruction in skin or corrosion of steel or aluminum. Examples include sulfuric acid and sodium hydroxide.

■ **Class 9 Miscellaneous hazardous materials** are a general grouping comprised of mostly hazardous waste. Dry ice, molten sulfur, and polymeric beads are examples that would use a class 9 placard, shown in **Figure 2-17**. This is known as a catch-all category: if a substance does not fit into any other category and presents a risk during transportation, it becomes Class 9.

■ **Other Placards and Labels**

NOTE A "Dangerous" placard is used when the shipper is sending a mixed load of hazardous materials if all of the materials are found in Table 2.

■ If the shipper sends 2,000 pounds of corrosives and 1,000 pounds of a flammable liq-

Figure 2-17 Class 9 placard and labels.

Figure 2-18 "Dangerous" placard.

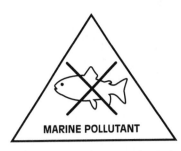

Figure 2-21 "Marine pollutant" placard.

Figure 2-19 "Chlorine" placard.

It is a subdivision that includes ammunition and consumer commodities, such as cases of hair spray. The package will have the printing "ORM-D" on the outside of the package. The previous regulation listed ORM-A through ORM-E, which are now grouped together in Class 9 Miscellaneous hazardous materials.

uid, then instead of displaying two placards the shipper can display a Dangerous placard, as shown in **Figure 2-18.** If any of the items exceeds 2,205 pounds picked up at one location, the shipper must display the placard for the material that exceeds 2,205 pounds in addition to the Dangerous placard.

■ Specific Name placards are sometime used instead of a generic class placard. A placard like the one shown in **Figure 2-19** with the name *chlorine* can be used in place of the poison gas placard, and an oxygen placard can be used instead of an oxidizer placard.

■ "Store Away from Foodstuff" is a placard that indicates a poisonous material is being transported, but is not poisonous to meet the rules to be placarded as a poison. Most chemicals in this category are PG-III. Chloroform is an example that would use the placard like the one shown in **Figure 2-20.**

■ ORM-D Other Regulated Material–Class D is left over from a previous DOT regulation.

■ "Marine Pollutant" is displayed on shipments that, if the material were released into a waterway, would harm marine life. The placard is shown in **Figure 2-21.**

■ Elevated temperature material will have a "HOT" label either to the side of or on the placard as shown in **Figure 2-22** if it meets one of the following criteria:

Liquid above 212°F
Liquid that is intentionally heated and has a flash point above 100°F
Solid at 464°F or above.

■ "Infectious Substances" is a label like the one shown in **Figure 2-23** that is sometimes used on the outside of trucks. It is not required by the DOT but may be required by other agencies, such as the Department of Health and Human Services or a state agency.

Figure 2-20 "Harmful" placard.

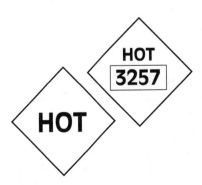

Figure 2-22 "Hot" placard.

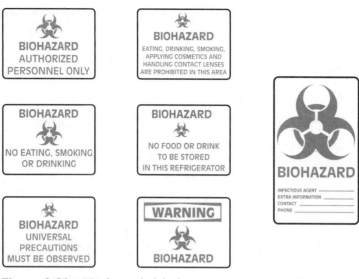

Figure 2-23 "Biohazard" labels.

■ The "Fumigated" placard, like the one shown in **Figure 2-24,** is used when a trailer or railcar has been fumigated with a poisonous material. This placard is most commonly seen near a port where containers are frequently fumigated after arriving from a foreign port.

■ "Residue" is a placard that is used in rail to identify the last product the tank held, as shown in **Figure 2-25.** Residue can be up to 3 percent of the total volume of the tank.

■ The white square background, as shown in **Figure 2-26,** is used to identify the following shipments while on rail: Explosives 1.1, 1.2, Poison gas or poison gas residue (Division 2.3 Hazard zone A). It is also used when controlled quantities of radioactive materials are transported by truck.

Problems with the Placarding System

The placarding system relies on a human to determine the extent of the load, determine the appropriate hazard classes, and interpret difficult regulations to determine if a placard is required. The placard then must be

Figure 2-25 "Residue" placard.

Figure 2-24 "Fumigation" label.

Figure 2-26 White square placard.

affixed to all four sides of the vehicle before shipment. Placards are only required for shipments that exceed 1,001 pounds, except for Table 1 materials, which require placarding at any amount. Authorities estimate that about 10 percent of the trucks traveling the highway are not placarded or not placarded correctly.

CAUTION Given the restrictions of many cities on bridges and tunnels where hazardous materials are not allowed, it can be anticipated that many trucks are not carrying the proper designations.

A placard can come off during transport and legally need not be immediately replaced. The fact that an incident involves a truck or train should alert the first responder to the potential for hazardous materials; when a placard is involved, extra precautions should be taken.

Labeling and Marking Specifics

Package markings must include a shipping name of the material, the UN/NA identification number, and the shipper and receiving companies' names and addresses. Packages that contain more than a **reportable quantity (rq)** of a material must also be marked with an *rq* near the shipping name. Packages that are listed as ORM-D materials should be marked as such. Some packages containing liquids must use orientation arrows. Materials that pose inhalation hazards are to affix an Inhalation Hazard label next to the shipping name as shown in **Figure 2-27.** Hazardous wastes will be marked waste or will use the EPA labeling system to identify these packages.

Labels are identical to placards, other than their size.

NOTE Materials that have more than one hazard may be required to display a primary hazard label and a subsidiary label.

The primary label will have the class and division number in the bottom triangle, while the subsidiary label will not have the number at all, as shown in **Figure 2-28.** As an example, the material acrylonitrile, inhibited, is required to be labeled flammable with a subsidiary label of poison.

OTHER IDENTIFICATION SYSTEMS

NFPA 704 System

It is designed for buildings, not transportation, and alerts the first responders to the potential hazards in and around a facility. The

NFPA 704
One of the other more common systems used to identify the presence of hazardous materials is the NFPA 704 system.

system is much like the placarding system and relies on a diamond-shaped sign divided into four areas, as shown in **Figure 2-29.** The four areas are divided by color as well, and use a ranking system to identify severity. The four areas and colors are:

■ Health hazard—Blue
■ Fire hazard—Red
■ Reactivity hazard—Yellow
■ Special hazards—White

CASE STUDY

A tractor-trailer was involved in an accident and had jackknifed in Baltimore County, Maryland. The HAZMAT team was called to assist with the fuel spill. Upon arrival, the HAZMAT team inquired as to the contents of the truck and were told by the first responders that the truck was carrying rocking chairs. While working to control the fuel leak, the HAZMAT team noticed a greenish-blue liquid coming from the front of the trailer, where it had been damaged in the accident. They located the driver and asked him about the contents of the trailer, which did not display a placard. The driver stated he was

carrying rocking chairs, which concurred with the shipping papers. The crew then opened the back of the trailer to inspect the cargo and found rocking chairs. Upon closer inspection of the front of the truck, the crew noticed several drums in the very front of the trailer. When the HAZMAT team and a state trooper confronted the driver, he confessed he had picked up a load of unknown waste from his first stop. He did not know the contents of the drums, and it took several days to determine the exact contents. It is unlikely that someone trying to evade the law will mark the shipment appropriately.

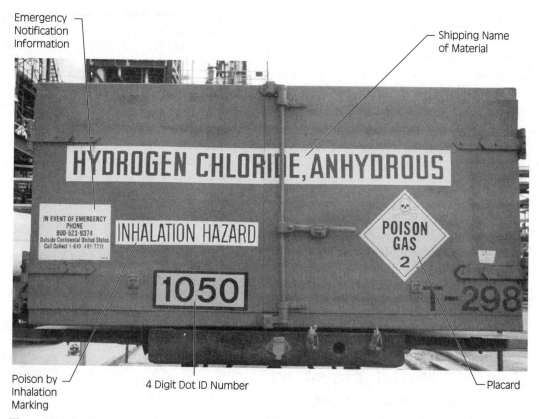

Emergency Notification Information

Shipping Name of Material

HYDROGEN CHLORIDE, ANHYDROUS

IN EVENT OF EMERGENCY
PHONE
800-523-9374
Outside Continental United States
Call Collect 1-610-481-7711

INHALATION HAZARD

POISON GAS 2

1050

T-298

Poison by Inhalation Marking

4 Digit Dot ID Number

Placard

Figure 2-27 The DOT adds the "Poison Inhalation Hazard" label to those materials that present severe toxic hazards.

The system uses a ranking of 0 to 4 with 0 presenting no risk, while a ranking of 4 indicates severe risk. The specific listing is as follows:

■ Health is based on a limited exposure to the materials using standard firefighting protective clothing as the protective clothing for the exposure.
  4—Severe health hazard
  3—Serious health hazard
  2—Moderate hazard
  1—Slight hazard
  0—No hazard

■ Flammability is the ability of the material to burn or be ignited.
  4—Flammable gases, volatile liquids, pyrophoric materials
  3—Ignites at room temperature
  2—Ignites when slightly heated
  1—Needs to be preheated to burn
  0—Will not burn

■ Reactivity is based on the material's ability to react, especially when shocked or placed under pressure.
  4—Can detonate or explode at normal conditions
  3—Can detonate or explode if strong initiating source is used
  2—Violent chemical change if temperature and pressure are elevated

POISON
6

CORROSIVE

Figure 2-28 The primary hazard is placed on top and has a class number on the bottom triangle. The subsidiary or secondary hazard is on the placard that is placed below the primary hazard placard. The subsidiary placard is also indicated by a lack of class number.

0

3   0

OXY

Figure 2-29 NFPA 704 placard.

1—Unstable if heated

0—Normally stable.

■ Special hazards are used to indicate water reactivity and oxidizer, which are included by the NFPA 704 system. In some cases other symbols may be used such as the tri-foil for radiation hazards, ALK for alkalis, and CORR for corrosives. The presence of a slashed W indicates an accompanying ranking structure for water reactivity in addition to the hazards listed in the other triangles:

4—Not used with the slashed W and a reactivity ranking of 4

3—Can react explosively with water

2—May react with water or form explosive mixtures with water.

1—May react vigorously with water

0—Slashed W is not used with a reactivity ranking of 0.

One potential problem with the NFPA 704 system is that it groups all of the chemical hazards listed in a building into one sign. If the sign is placed on a tank that contains one material, the system does a good job of warning about the contents of the tank, but it does not provide the name of the product. For a facility with hundreds if not thousands of materials, the system will only warn of the worst-case scenario. As an example, dramatically different tactics are used to handle a flammable gas incident versus a flammable liquids incident, but both can be classified as fire hazard 4. The system is best used to alert the first responder as to the presence of hazardous materials and warn of the worst-case scenario.

Hazardous Materials Information System

Commonly referred to as HMIS, the Hazardous Materials Information System was designed to provide a mechanism to comply with the Hazard Communication regulation. This system may also be known as Hazardous Materials Information Guide (HMIG). The Hazard Communication regulation requires that all containers be marked with the appropriate hazard warnings and the ingredients be provided on the label, like those shown in **Figure 2-30.** Many products that come into the workplace are missing adequate warning labels. The HMIS is not a uniform system, but can be developed by the facility or by the manufacturer of the labels. Thus, one system may vary from another. Most systems are similar to the NFPA 704 system and use blue, red, and yellow colors with a numbering system that provides an indication of hazard. The colors may be used in a triangle format or, in most cases, in stacked bars.

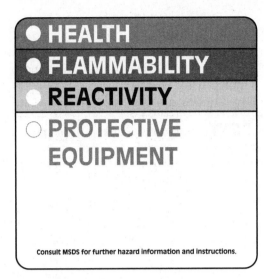

Figure 2-30 HMIS label.

The numbers are usually 0–4, the same as the NFPA system, but in rare cases may differ from the NFPA system. The facility manager or other representative should have the key to the symbols, or there should be a chart indicating the symbols and warning levels somewhere in the facility, usually with the Material Safety Data Sheet (MSDS). In most cases the MSDS and other hazard communication information should be in a central location. Some systems include a picture of the level of the personal protective equipment (PPE) that is required for the substance.

Each HMIS system is different, and responders should not assume what the actual hazard level is until warning levels can be determined. When conducting inspections or preincident walk-throughs, responders should determine the location of the MSDS information and make sure it is accessible during emergencies. In some cases the MSDSs may be stored in or near the chemical storage area, which during an emergency may be inaccessible. Responders should suggest that another set be provided at a guardhouse or in a locked box on the outside of the building. One large industrial complex with a large number of individual labs had a fire in one of these labs. The responding firefighters correctly asked for the MSDSs for the materials in the lab, only to be told that they were in the lab. The MSDSs were placed inside the lab for the employees' use, and their location was convenient for that purpose. After the fire emergency responders requested that the plant store a master copy of all MSDSs in the security office and move the MSDSs for the individual labs to the outside of the lab. MSDS boxes were placed outside each lab area, so that employees would have easy access and, in the event of an emergency, firefighters could also reach the MSDS for that lab without having to enter the lab.

Military Warning System

Whenever possible, the military uses the DOT placarding system, but in some cases it may employ its own system.

SAFETY Within the DOT's North American Emergency Response Guidebook is an emergency contact number for responding to an incident involving a military shipment.

In most cases for extremely hazardous materials, arms, explosives, or secret shipments, it would be safe for emergency responders to assume prior to their arrival that the military is aware of the incident and probably is responding. The higher the hazard, the more likely there will be an escort for the shipment. There may be shipments in which the driver of the truck is not allowed to leave the cab of the truck, and may provide warnings to stay away from the truck.

SAFETY On high security shipments the driver and the personnel in the escort vehicle are armed. If you have an incident involving one if these vehicles, obey the commands of the escorting personnel and determine if they have made the appropriate notifications.

If the driver and escort crew are killed or seriously injured in the accident, it would be advisable to notify the military about the incident, although with satellite tracking, help is probably on the way. The phone number to contact the military is in the DOT NAERG, along with **Chemtrec's** and other emergency contact numbers. Other incidents involving fuels, food, or military equipment may require you notifying the military of the incident. The military typically uses its own marking system at its facilities to mark the buildings. Military facilities use a series of symbols and a numerical ranking system shown in **Figure 2-31.**

Pipeline Markings

At any place an underground pipeline crosses a mode of transportation, the pipeline owner is required to place a sign like the one shown in **Figure 2-32** that indicates:

■ Pipeline contents
■ Pipeline owner
■ Pipeline emergency contact number.

The pipeline contents may be general as in "petroleum products" as the pipeline may ship fuel oil, gasoline, motor oil, or other products in the same pipe. Dedicated pipelines, which only carry one product, will be marked with that specific product.

Figure 2-31 Military placarding system.

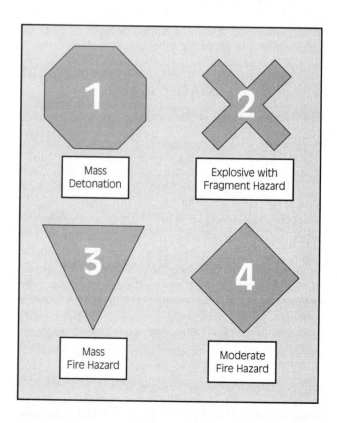

Contents of Pipeline

Owner

Phone Number

Figure 2-32 The owner of a pipeline is required to provide the contents of the pipe, the owner name, and an emergency contact number.

The pipeline should be buried a minimum of 3 feet and should be adequately marked. Many larger pipelines, such as the Colonial pipeline, originate in Texas and end in New York; the Colonial is 26 inches in diameter along the main pipeline. In the event of a release involving the pipeline the line will be immediately shut down. Even with this immediate shut down, there is the potential to lose several hundred thousand gallons of hazardous materials if the distance to the shutoff is substantial. An incident involving a pipeline can be a serious event; do not underestimate the need for considerable local, state, and federal resources. Just within the responders considerable resources may be required, such as command staff, logistic support, communications, and tactical units. A fuel oil pipeline rupture in Reston, Virginia, resulted in the loss of more than 400,000 gallons of fuel, requiring considerable resources from several states to control the spill. The resources included emergency response organizations, local, state, and federal assistance, and a considerable number of private cleanup companies.

NOTE Some pipelines move one type of product, while others move several different types each day.

The product in the pipelines varies from liquefied gases and petroleum products to slurried material.

Pipeline companies are required to conduct in-service training and tours for emergency responders in the communities their pipelines transverse. When you have an incident on or near any pipeline, it is advisable to notify the pipeline owner of the incident, even if you are fairly certain the pipeline was not damaged. A train derailment in California caused a pipeline to shift, and the pipeline did not release any product until several days after the original derailment occurred. Most pipeline operators would like to have the opportunity to check the line to head off the potential for a catastrophic release several days later. After several catastrophic pipeline releases the pipeline industry has been required to establish contact with local responders. This contact usually involves a tour of pumping stations, storage facilities, and an overview of the operator's system.

Container Markings

Most containers such as drums are marked with the contents of the drum, like the one shown in **Figure 2-33.** Cylinders should have the name of the product stenciled on their side. In bulk shipments the

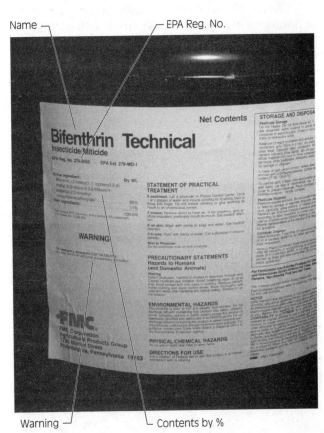

Name

EPA Reg. No.

Warning

Contents by %

Figure 2-33 The label describes the contents of the drum and provides precautions to be used when the product is handled. The EPA signal word is a significant clue to the hazards of the product.

bulk container will have the name of the product stenciled on the side. In bulk shipments the DOT requires that some products have their own placard, such as the chlorine placard. Trucks that are dedicated haulers will also stencil or mark the product's name on four sides of the vehicle.

Pesticide Container Markings

Due to their toxicity, required markings for pesticides are regulated by the EPA. The label on a pesticide container such as the one in **Figure 2-34** will have the trade name of the pesticide, which is chosen by the manufacturer and is not usually the chemical name for the product.

NOTE The pesticide label will also contain a signal word: Danger, Warning, or Caution.

In the United States the EPA issues an EPA registration number and in Canada the label will have a pest control number. The label will also include a precautionary statement and a hazard statement; examples are: "Keep from waterways," and "Keep away from children." The active ingredients will be listed by name and percentage, which in most cases for active ingredients are a small percentage of the product. Inert ingredients are also listed but not specifically named. For liquid pesticides the inert ingredient is usually kerosene or diesel fuel, an item one does not normally consider inert, except under these EPA guidelines.

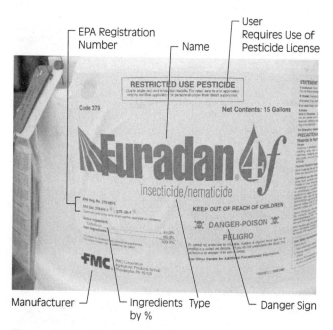

Figure 2-34   Pesticide label.

CONTAINERS

General Containers

Hazardous materials come in a variety of containers of many shapes and sizes, from 1-ounce bottles to tanks to ships carrying hundreds of thousands of gallons. Considering the materials you have in your home, you see a wide variety of storage containers. You have compressed gas cylinders holding propane, steel containers holding flammable and combustible liquids, and bottles, jars, and small drums holding various products. You can also locate cardboard boxes, plastic lined cardboard boxes, and bags of various types. The type of material and the end use for the product determines the packaging. Packaging used to store household or consumer commodities is usually different than the industrial version. In some cases the industrial version may be full-strength undiluted material, while the household version is only a small percentage of that strength mixed with a less hazardous substance such as water. The type of container is usually a good clue as to the contents of the package.

The more substantial, durable, fortified a container is, like the container shown in **Figure 2-35,** the more likely the material inside is dangerous. On the other hand, materials that are transported in fiberboard drums are usually not of significance in regard to human health, although they may pose a risk to the environment.

During the recognition and identification process, first responders should be alert for anything unusual when arriving at an incident. When on an EMS call to a residential home, it would be unusual to find a 55-gallon drum in a bedroom along with glassware associated with a lab environment. These types of recognition and identification clues should alert the first responders that additional assistance may be required. Arriving at an auto repair garage and finding 55-gallon drums and compressed gas cylinders should not be unexpected, however.

Cardboard Boxes

With the popularity of shopping clubs and discount warehouses, more and more homeowners are purchasing cases of materials, when in the past they only bought a gallon or less. Cardboard boxes are one method of shipping and containing hazardous materials. They can hold glass, metal, or plastic bottles. In some cases they may have a plastic lining such as the box shown in **Figure 2-36,** which holds sulfuric acid. Many household pesticides, insecticides, and fertilizers are contained in cardboard boxes. With the exception of these products and the sulfuric acid shown in

Figure 2-35 The type of container can offer some clues as to its contents. As this drum is reinforced, it has a high likelihood of containing an extremely hazardous material.

Figure 2-36, most products contained in boxes are usually not extremely toxic to humans, but they may present an environmental threat. Materials in transport to suppliers may be transported in larger cardboard boxes and then broken down at the retail level. Responders should note any labels on these packages, but the absence of any labels does not indicate that hazardous materials are not present.

BOTTLES

From 1-ounce bottles to 1-gallon bottles, the variety is endless, and the types of products that are contained are too numerous to mention. In recent times manufacturers have taken care in packaging their materials for safe transport and use. In the years past, most chemicals were shipped in glass bottles. When shipped, these bottles are usually packed in cardboard boxes and are insulated from potential damage. One-gallon glass containers are usually shipped in what is known as carboys, like those shown in **Figure 2-37.** If the container were dropped, these protective covers should ensure that the bottle will survive

Figure 2-36 Typically chemicals that can cause harm are not packaged in cardboard. This sulfuric acid is one example of a product that can cause harm, and is in cardboard

the fall and prevent any potential damage during transportation. Carboys are usually seen in laboratories and in smaller chemical production facilities.

Ensuring compatibility with the container is important but the one area that usually results in a release is the use of an improper cap. The chemical

Figure 2-37 To protect the glass bottle containing a corrosive, a carboy is used to protect the bottle if dropped.

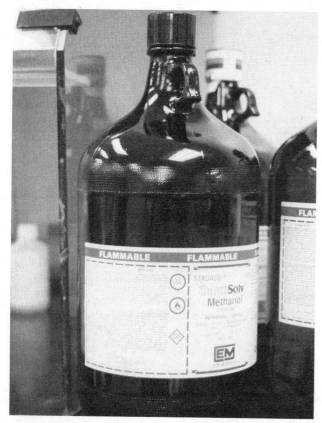

Figure 2-38 This glass jar is coated with plastic that will not allow the liquid to spill out if the glass is broken or dropped.

Figure 2-39 It can be quite surprising to open the back of a tractor-trailer and find these super sacks. They contain solid materials, some of which may be toxic.

must be compatible with glass and the material used to construct the lid. There are a variety of materials that are used in the manufacture of lids. Like the example in **Figure 2-38,** many new glass containers are coated with plastic to keep the bottle from breaking when dropped. Even if the bottle is cracked, the contents should remain sealed within the plastic coating.

BAGS

From paper bags to super sacks that weigh several tons, bags also contain a variety of materials. Bags can be simple paper bags, plastic-lined paper bags, fiber bags, plastic bags, or reinforced super sacks or tote bags. Even liquids are being transported in containers called bladders, which mimic the super sacks. It can be a surprise to open the back of a trailer and find four super sacks like those shown in **Figure 2-39** carrying a material that is classified as a poison. Bags carry anything from food items to poisonous pesticides, and the method of transportation varies widely.

DRUMS

When discussing hazardous materials, these are the containers most responders call to mind. They vary in

size from 1 gallon up to a 95-gallon overpack drum. The construction varies from fiberboard to stainless steel. The typical drum contains 55 gallons and weighs 400 to 1,000 pounds. It is possible to get an idea of what a drum may contain by the construction of the drum. **Table 2-6** provides an indication of potential drum contents, but this is not an absolute listing. Contents can and do vary drum to drum.

CYLINDERS

Cylinders like those shown in **Figure 2-40** come in 1 pound sizes up to several thousand pounds and carry a variety of chemical products. The product and its chemical and physical properties will determine what type of container the product is stored in.

CAUTION Other than the hazard of the chemical, a major concern that all cylinders pose is the fact that they are pressurized.

The pressures range from a low of 200 psi to a high of 5,000 psi. One of the most common cylinder firefighters run across is the propane tank, found in 1 pound up to millions of gallons. When used in residential homes, you will find the 20-pound cylinder for barbecues and cylinders of 100 to 250 pounds used as a fuel source for the home. In some areas it is not uncommon to find 1,000-gallon cylinders and many times now they are being buried underground.

Some specialized cylinders hold cryogenic (extremely cold) gases that appear to be high pressure but in reality are low pressure. The bulkiness of the cylinders is due to the large amount of insu-

TABLE 2-6

Drum Contents	
Type of drum	**Possible contents (in order of likelihood)**
Fiberboard (cardboard) unlined	Dry, granular material such as floor sweep, sawdust, fertilizers, plastic pellets, grain, etc.
Fiberboard—plastic lined	Wetted material, slurries, foodstuffs, material that may affect cardboard or could permeate the cardboard
Plastic (poly)	Corrosives such as hydrochloric acid and sodium hydroxide, some combustibles, foodstuffs such as pig intestines
Steel	Flammable materials such as methyl alcohol, combustible materials such as fuel oil, motor oil, mild corrosives, and liquid materials used in food production
Stainless steel	More hazardous corrosives such as oleum (concentrated sulfuric acid)
Aluminum	Pesticides or materials that react with steel and cannot be shipped in a poly drum

Figure 2-40 Cylinders present additional risks to responders. Not only can the contents be hazardous, but if the cylinder is involved in a fire, it may explode.

lation required to keep the material cold. Cylinders usually have **relief valves** or **frangible disks** in the event they are overpressurized or are involved in a fire. Communities of all sizes have cylinders of chlorine and sulfur dioxide used in water treatment. These cylinders come in 100- and 150-pound and 1-ton cylinders, which could create a major incident if they were ruptured or suffered a release.

NOTE The area affected by a 100-pound chlorine cylinder release can be several miles, causing serious injuries if not fatalities. **Figure 2-41** depicts the vapor cloud that could result from such a release.

Totes and Bulk Tanks

Both **totes** and **bulk tanks** are becoming more common as they are sized between drums and tank trucks; they are used for a variety of purposes. They hold flammable, combustible, toxic, and corrosive materials. Totes and bulk tanks are used in industrial and food applications. They are constructed of steel, aluminum, stainless steel, lined materials, poly tanks, and other products, such as those shown in **Figure 2-42.** They carry up to 500 gallons; a typical size is 300 gallons. They are transported on flatbed tractor-trailers or in box-type tractor-trailers. A common HAZMAT incident involving totes occurs when tanks are

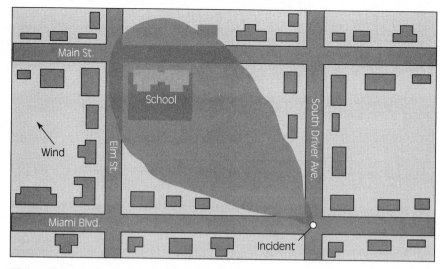

Figure 2-41 The type of vapor cloud commonly referred to as a plume varies with the terrain and buildings in the vicinity. The plume here represents one of the most common types.

Figure 2-42 The use of 55-gallon drums is decreasing and the use of these portable bulk tanks is increasing. Like super sacks, these can be hidden away in the back of a trailer.

loaded off the bottom through a swinging valve, such as the one shown in **Figure 2-43**. It is a common occurrence for this valve to swing out during transport and get knocked off during movement, as shown in **Figure 2-44**. An unusual tote is the type designed to transport calcium carbide, a material that, when it gets wet, forms acetylene gas, which is reactive and very flammable. A calcium carbide tote is shown in **Figure 2-45**.

Figure 2-43 The most common type of spill occurs when this valve is knocked off, releasing the contents.

Figure 2-44 The tote in the middle is the nitrocellulose tote. The spill emptied the contents of the tote and covered the floor of the warehouse. On top of the spill is a layer of foam, which did not reduce the flammable vapors. The LEL readings were 90 percent of the LEL at some points during the operation, a very dangerous situation. Complicating the release is the fact that nitrocellulose becomes very shock sensitive as it dries and can detonate. This photo shows numerous other totes containing nitrocellulose, and most of the drums contain flammable waste. It was dangerous to operate in this environment, but the resulting drying was even more dangerous so speed was necessary to reduce the potential for explosion. *(Photo by the Maryland Department of the Environment)*

Figure 2-45 Being carried on this railcar are specialized totes used to carry calcium carbide. When transported, calcium carbide is placarded "Dangerous when wet." The perception is that the addition of water to a material placarded in this fashion would cause an explosion. In the case of calcium carbide the addition of water would result in the production of acetylene gas, which is very flammable and would present a fire hazard. *(Photo by Greg Socks)*

Pipelines

Pipelines vary in size and pressure. They are typically from ½ inch to in excess of 6 feet in diameter and are commonly buried underground. The most common products are natural gas, propane, and assorted liquid petroleum products. The larger petroleum pipelines originate in Texas and Louisiana and proceed up throughout the East Coast. The West Coast also has its share of larger pipelines, with a large majority originating in Alaska. Pipelines can originate from any bulk storage facility and can cross many states. Some type of pipeline system is found in every state. Since the amount in the pipelines varies, it is important that first responders know the location of the pipelines and emergency contact names and phone numbers so if there is a suspected problem they can notify the pipeline owner immediately.

Highway Transportation Containers

The type of vehicle provides some important clues as to the possible contents of the vehicle. Four basic tank truck types carry hazardous materials, with some additional specialized containers. The most common truck involved in chemical releases are tractor-trailers or box trucks, based upon the high numbers of trucks traveling the highways. Tractor-trailers and box trucks carry the whole variety of hazardous materials and portable containers. They can carry loose material that is not contained in any fashion other than the truck itself. They can carry portable tanks that hold 500 gallons or bulk bags that weigh several tons.

CAUTION When dealing with tractor-trailers, the rule is: Expect the unexpected.

Nothing is routine. Until you interview the driver, review the shipping papers, and finally examine the cargo, you cannot confirm or deny any hazardous materials. Sometimes the signage on the trailer is an indication of the possible contents, and a trailer that has several placard holders is a likely candidate for transporting hazardous materials. If a tractor-trailer is refrigerated, like the one shown in **Figure 2-46,** and is carrying hazardous materials, these require extra precautions as they may require the cold temperature to remain stable.

Figure 2-46 Although in most cases refrigerated trailers are carrying food, there exists the possibility that they may have chemicals that require refrigeration to remain stable.

Figure 2-47 Many of the differences between a DOT-306 and a 406 are internal. The biggest difference is that the dome covers on a 406 are less likely to open during a rollover, although the skin of the tank is thinner. *(Courtesy of Maryland Department of the Environment)*

A common box trailer that is involved with leaking containers is known as an **intermodal container,** more commonly called a **sea container.** Throughout the United States these containers are used for transport of materials. Although common in and around marine terminals, these containers can be found anywhere. They are typically used on ships, then offloaded onto a tractor-trailer or loaded directly onto a flatbed railcar. These containers come from all over the world and can contain any imaginable commodity. The types of containers that are shipped in these trailers may be bags, boxes, drums, bulk tanks, or cylinders. The reasons they are involved in many incidents are the number of containers on the highway and the many stresses subjected on these materials during transport. Although the driver is supposed to have the shipping papers for the contents, on occasion the paperwork is missing or is sealed in the back of the trailer. Finding out the contents of the trailer can be very difficult and frustrating.

Tank trucks carry several hundred gallons up to a maximum of 10,000 gallons. The DOT allows maximum loads by weight not by gallons, so the actual gallonage varies state to state. The most common tank truck is the gasoline tank truck, which usually carries 5,000 to 10,000 gallons. The DOT changed its regulations in 1995 covering tank trucks, and you may see two systems for identifying tank trucks, like those shown in **Figure 2-47.** In the past the DOT wrote specifications as to how the manufacturer would build a tank truck. Today, the federal agency has established performance-based standards for vehicle construction. The DOT allows trucks that were manufactured before the new regulation to remain on the road, as long as they meet the applicable inspection requirements. The four basic type of tank trucks are:

- DOT-406/MC-306 Gasoline tank truck[2]
- DOT-407/MC-307 Chemical hauler
- DOT-412/MC-312 Corrosive tanker
- MC-331 Pressurized tanker

The DOT numbers are the more recent manufactured tanks, and the MC (motor carrier) numbers identify those tanks manufactured prior to September 1995. There are some differences in construction of the two types, some in favor of emergency responders, others in favor of lower cost to the shipping company, but overall the newer tanks hold up better during accidents and rollovers. If you are not sure of the type of tank, all tank trucks have **specification (spec) plates** that list all the pertinent information. The spec plate is located on the passenger side of the tank near the front of the tank. In some cases, you may find shipping papers or material safety data sheets in a paper holder as shown in **Figure 2-48.**

Figure 2-48 In most cases the shipping papers will be with the driver in the cab of the truck, but they may be in a special tube located on the trailer.

DOT-406/MC-306

This is the most common tank truck on the road today, as shown in **Figure 2-49** and **Figure 2-50.** These trucks are involved in the most accidents mainly due to the sheer number of shipments and trucks on the road. Although the most common products carried on these trucks are gasoline and diesel fuel, almost any flammable or combustible liquid can be found on these types of trucks. This truck is known for its elliptical shape and is usually made of aluminum. The maximum pressure that this truck can hold is 3 psi.

The trucks generally have three or four separate tanks called pots, but two to five tanks are not uncommon. The valving and piping is contained on the bottom of the tank, as shown in **Figure 2-51,** and the number of pots is indicated by the number of outlets and the number of manhole assemblies located on top of the tank. The pots are separated by bulkheads, which tend to fail during a rollover situation.

Figure 2-49 The MC-306/DOT-406 is the most common truck on the highway today and is referred to as a gasoline tanker. It can and frequently does carry other types of flammable and combustible liquids.

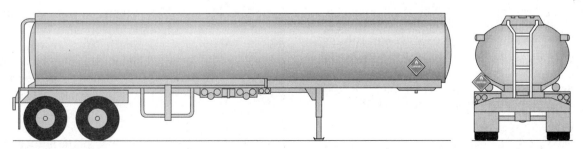

Figure 2-50 The silhouette of an MC-306/DOT-406 is known as being elliptical from the ends and circular from the sides. There is no way of knowing the difference between a 306 and a 406 from a distance, and the best way to confirm the construction type is to check the specification plate found near the forward running gear.

Figure 2-51 Most 306/406 trucks have more than one tank (pot), and the number of tanks is indicated by the number of valves or the number of dome covers on top of the truck. *(Courtesy of Maryland Department of the Environment)*

Although in most states the shipper is not supposed to have a mixed load of flammable and combustible materials, responders may encounter a widely differing load pot to pot.

> **CAUTION** During a rollover, the bulkheads on a tank truck may shift, allowing all the products to mix and form one big tank.

By an initial examination it may appear only one pot is ruptured and leaking, but it is possible to lose the whole contents of the tank through a leak in one pot, as shown in **Figure 2-52**. In the majority of rollovers, experience has shown that at least one bulkhead has separated in almost every accident, resulting in product mixing. In most cases this is not a serious problem as it may be the mixing of grades of gasoline. Within the individual tanks are baffles that limit the movement of the product within the tank.

The emergency shutoff valves, as shown in **Figure 2-53,** are located on the driver's side near the front of the tank, and one is located near the piping on the passenger side. Newer style tanks have considerable vapor recovery systems and rollover protection; in some cases these are combined. Older style tanks used to be made of steel, which presented an explosive situation known as a **boiling liquid expanding vapor explosion (BLEVE).**

DOT-407/MC-307

These are the workhorses of the chemical industry, as shown in **Figure 2-54.** They carry a variety of materials, including flammable, combustible, corrosive,

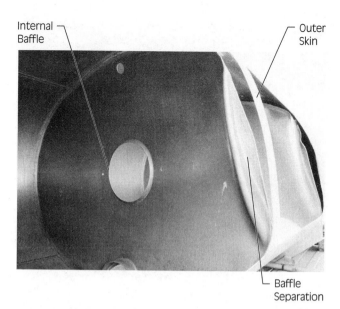

Internal Baffle — Outer Skin — Baffle Separation

Figure 2-52 When a truck rolls over on its side, the bulkheads and baffles usually shift, allowing chemicals to mix throughout the tank. Bulkheads separate the various tanks within the truck, while baffles limit the movement of the product within the individual tank. During a rollover these will separate from the sides of the tank. Shown in the photo where most of the sides of the truck have been cut away, you can see the separation of the baffle (hole in the middle indicates a baffle) from the side wall of the tank.

poisonous, and food products. There are two basic types of chemical tanks, one that is insulated and one that is not. The insulated tank can have a number of additional concerns that do not apply to an uninsulated tank (see **Figure 2-55** and **Figure 2-56**). These

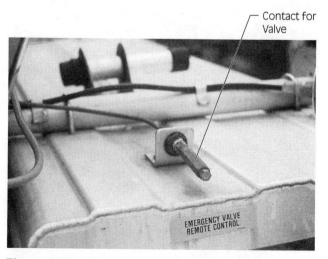

Contact for Valve

Figure 2-53 Most trucks have at least one emergency shutoff, and some have two. The most common location is near the driver's door, and the other is usually located near the valve area.

Figure 2-54 This is an uninsulated MC-307 tank on the right. It has reinforced stiffening rings and is a single tank. It does not have separate compartments.

tanks usually hold 2,000 to 7,000 gallons, lower amounts as compared to the 306/406 trucks because most of the products they carry are heavier than petroleum products. The average amount found in these tanks is 5,000 gallons.

The uninsulated tank is round and has stiffening rings around the tank, and the offloading piping is located on the bottom or off the rear of the tank, as shown in **Figure 2-57** and **Figure 2-58.** The loading piping and manhole is usually on the top in the middle. These tanks are comprised of only one pot and do not have separate bulkheads or pots. These tanks generally do not hold up as well as the insulated version during rollovers and accidents. The shell is made of stainless steel and can hold pressures up to 40 psi.

The insulated tank is a covered version of the uninsulated version, although in some cases it is a slightly smaller inner tank. The inner tank is made

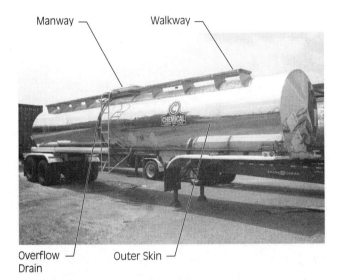

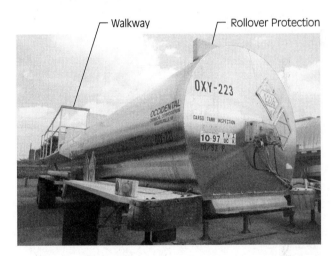

Figure 2-55 The insulated version of the tank is identical to the uninsulated version but with an aluminum cover and insulation. Note the difference with these two insulated tanks. The one on the right has safety railings around the manway. Although not an absolute rule, the truck on the right tends to carry a more toxic product and has added safety features. Most are not required but were added by the trucking company.

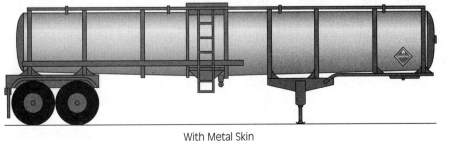

With Metal Skin

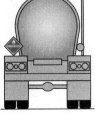

Horseshoe Shape from Rear

Figure 2-56 The silhouette of an insulated MC-307/DOT-407 on the right is horseshoe shaped from the front and rear and circular from the side. The truck on the left is an MC-307/DOT-407 non insulated.

Figure 2-57 This is an uninsulated MC-307/DOT-407, commonly referred to as a chemical hauler. *(Photo by Greg Socks)*

Hookup Lines for Heating or Cooling

Hookup Lines for Heating or Cooling

Figure 2-59 Products carried in an insulated 307/407 either need to remain heated or cooled. Some products may need to be heated via heater coils, which run around the tank, before they can be offloaded.

of stainless steel with about 6 inches of insulation, and the outer shell is made of aluminum. The inner tank may also be lined with fiberglass or other liner depending on the chemical that is carried. Due to the aluminum outer shell and insulation, these tanks hold up remarkably well during rollovers.

SAFETY One of the major problems with this insulated tank is that, in the event of a leak, the point where the material leaks out of the outer shell is usually nowhere near the leak on the inner shell.

Within the insulation there can be heating and cooling lines, depending on the product being carried, such as those shown in **Figure 2-59.** Products such as paint are shipped at 170°F and need to be heated to that temperature to offload the paint. Some products need to remain at certain temperatures to remain stable, and first responders need to

be aware of any special requirements.

In general, both insulated and uninsulated tanks have rollover protection, similar piping, and relief valves that serve two purposes: overpressurization and vacuum protection. The emergency shutoffs are located near the front of the tank on the driver's side and near the offloading piping.

DOT-412/MC-312

Tankers like those shown in **Figure 2-60** and **Figure 2-61** carry a wide variety of corrosives, both acids and bases. These tankers are round and smaller than the 306s and 307s, due to the weight of the corrosives they carry. Most petroleum products weigh about 8 pounds per gallon while some corrosives weigh up to 15 pounds per gallon. Because of the

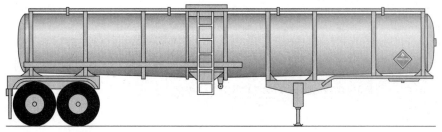

Without Metal Skin

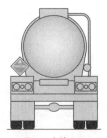

Round Shape from Rear

Figure 2-58 The silhouette of an uninsulated MC-307/DOT-407 is circular from the side and the rear. In some cases the side view may present what is known as a double conical, which can be described as two ice cream cones butted together to form the tank with the thickest part in the middle.

Figure 2-60 The MC-312/DOT-412 is designed to carry corrosives. Its design is similar to the uninsulated 307, although smaller. The inner tank may be lined with a variety of materials to prevent the corrosive from attacking the tank.

Protection Ring — Fill Point — Manway
Offload Pipe

Figure 2-62 One indicator that a truck is carrying a corrosive is the ring that circles a tank around the manhole. It is used to protect the tank from spillage.

weight the stiffening rings that are used are generally bulkier than the ones used on DOT-407 tanks. These tankers are constructed of a single tank that carries up to 7,000 gallons, with most tanks holding 5,000 or less. The tanks are made of stainless steel and are usually lined to protect against corrosion. The piping can be on top of the tank or in the middle but is usually located on the end of the tanker. The piping is usually contained within a housing that has the manhole and offload piping; this housing protects the piping in the event of a rollover, as shown in **Figure 2-62.** The area around the manhole is usually coated with a material, usually a black, tarlike coating, that resists the chemical being carried.

DOT-331

These tanks look like bullets and are noted for their rounded ends and smooth exteriors as shown in **Figure 2-63** and **Figure 2-64.** They carry gases that are liquefied by pressure; one of the most common

products carried in these tanks is propane. These tanks carry a wide variety of liquefied gases, such as ammonia, butane, and other flammable and corrosive chemicals. These tanks carry up to 11,500 gallons and have a general pressure of 200 psi, although it can be as high as 500 psi.

The relief valves are located on top of the tank at the rear of the trailer and can be damaged during a rollover. There is a possibility during a rollover that the relief valves will fail to function or, once activated, may malfunction. The steel tanks are uninsulated and heavily fortified with heavy bolts used in the piping and manholes. The tanks are usually painted white or silver to reduce potential heating by sunlight. The tanks contain a liquid along with a certain amount of vapor. The maximum allowable

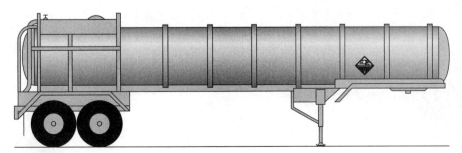

Figure 2-61 The silhouette of an MC-312/DOT-412 is circular from the side and the rear and has a smaller profile than an uninsulated MC-307/DOT-407. The tank hauls corrosive materials, which can weigh upwards of 15 pounds per gallon as compared to approximately 8 pounds per gallon for fuels.

Leve
Indicator

Control
Panel

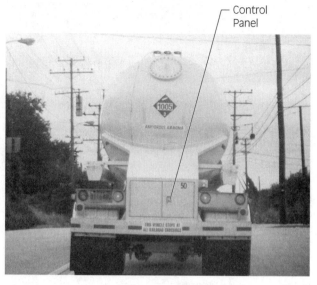

Figure 2-63 MC-331 tanks carry liquefied compressed gases such as propane and ammonia. They are made of steel and carry a variety of products.

liquid is 80 percent to allow for expansion when heated. The liquid in the tanks is at atmospheric temperature but upon release can go below 0°F and can even frostbite upon contact. The pressure in these tanks is of concern when responding to incidents involving these trucks.

When these tanks are involved in fire situations, a BLEVE may result in which both the heat and the pressure in the tanks increase.

SAFETY If the pressure increases at a rate higher than the relief valve can handle, the tank will explode. These explosions have been known to send pieces of the tank up to a mile away. Typically the ends of the tank travel the furthest, although any part is subject to becoming a projectile.

A large number of responders have been killed by propane tank BLEVEs; when a BLEVE such as the one shown in **Figure 2-65** occurs, it usually results in more than one firefighter's death.

Specialized Tank Trucks

MC-338 CRYOGENIC TANKERS

These tankers are a specialized construction, as the example in **Figure 2-66** shows. They are a tank with an outer shell. The inner container is steel or nickel, with a substantial layer of insulation; the exterior is aluminum or mild steel. The space between the shells is placed under a vacuum to serve as an insulation barrier. The ends of the tank are flat, and the piping is contained usually at the end of the tanker in a double door box. Relief valves are located on top of the tank to the rear of the tank. The best way to describe this tanker is to

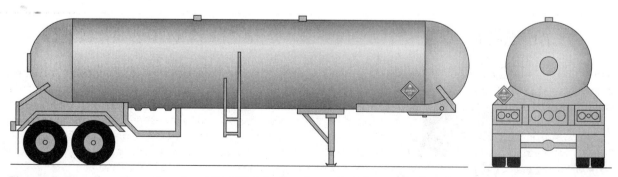

Figure 2-64 The silhouette of an MC-331 is bullet shaped, with rounded ends.

Fire impinging on a propane tank.

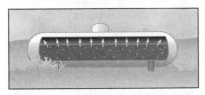

As heat increases inside the tank, the pressure also increases. The liquid will begin to boil.

As the pressure increases, the relief valve will open, releasing propane. The propane, being heavier than air, will sink.

The vapor will reach the fire and ignite.

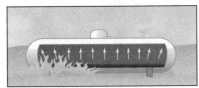

The relief valve will ignite, also causing heat to be increased on the tank by that flame. The pressure will increase in the tank.

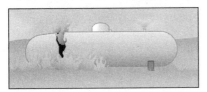

As the pressure in the tank is increasing, the tank may discolor and the pitch of the relief valve will get higher. Eventually the tank will rupture and the contents will ignite explosively. This is known as a BLEVE.

Figure 2-65   Diagram of a BLEVE.

TEMPERATURES AND PRESSURIZED CONTAINERS

When a propane tank is emptied, the temperature of the propane drops below 0°F as the pressure is reduced. Temperature, pressure, and volume are interrelated. Think of your SCBA bottle: When it gets filled, it becomes hot as you are increasing the pressure and volume. When you use the SCBA bottle, the bottle becomes cold, as it is losing pressure and volume. Any time you change one of the parameters, there is a corresponding change in the other properties. When a gas is pressurized to a point that it becomes a liquid, as is the case with propane, it allows for a lot of propane to be stored in a small container. When released, however, the temperature will drop as the pressure is decreasing in the container. This is known as *auto-refrigeration.*

Figure 2-66   The MC-338 carries refrigerated liquefied gases. The tank resembles a rolling Thermos™ bottle.

compare it to a Thermos™ bottle on wheels. Cryogenic materials are gases that have been compressed and cooled to a temperature that converts them to liquids. Unlike liquefied gases, which use pressure to reduce them to liquids, cryogenics are pressurized and cooled to the point of becoming liquid. To remain liquids the material must be kept cool. To be a cryogenic material the liquid must be at least −150°F and can be as cold as −456°F. The most common products are nitrogen, carbon dioxide, oxygen (liquid oxygen or LOX), argon, and hydrogen. As the material inside is kept liquid by the cooling effect and when on the road, the tank has a maximum of 25 psi. As the truck travels, the sun heats the material and the pressure increases. As the pressure increases, the relief valve will open up, relieving any excess pressure into the atmosphere.

NOTE It is not uncommon to see a white vapor cloud like that shown in **Figure 2-67** coming from the relief valves while the truck is traveling on the highway or sitting alongside the road. This is a normal occurrence and is not reason for alarm.

When the truck makes a delivery, the pressure must be increased to push the liquid out of the tank. The driver opens the piping and allows the material to flow into an evaporator, which is located just in front of the rear wheels; once in the evaporator the material will heat up, increasing the pressure in the tank and forcing the liquid into the receiving tank. During transportation and offloading a large amount of ice will be built up on the piping and valves.

TUBE TRAILERS

These tube trailers contain several pressurized vessels, constructed much like the MC-331 tank shown in **Figure 2-68.** They are constructed of steel and have pressures ranging from 2,000 to 6,000 psi. They hold pressurized gases such as air, helium, oxygen, and other gases. The piping and controls are located usually on the rear of the trailer, but could be in the front. The typical deliv-

Figure 2-67   When transporting liquids, it is not uncommon to see vapors coming from the relief valves. The truck can only travel with the tank at 25 psi or less. As it heats, the pressure increases, triggering the relief valve. This is a normal situation, and the tank will vent until below the 25 psi. When offloading, the pressure can be increased to assist with the offloading procedure.

Figure 2-68 The tubes on this trailer contain pressurized gases. The pressure can vary from 2,000 to 6,000 psi.

ery mode is that the driver will drop a full trailer off at a facility and pick up the empty one for refilling. Although they are not subject to a BLEVE because they only contain a gas, if involved in a fire these trailers can have what is known as a violent tank rupture (VTR) and rocket in the same fashion as a BLEVE.

DRY BULK TANKS

These slightly resemble large uninsulated MC-307s in shape with bottom hoppers to unload the product, as shown in **Figure 2-69.** The tanks hold dry prod-

ucts and sometimes a slurry, like concrete. The most common products are fertilizers, flour, grain, lime, and other food products. A potential hazard when dealing with these tankers is predominately environmental, but these tankers may contain toxic materials. They are usually offloaded using air pressure either from the truck itself or at the facility.

HOT MATERIALS TANKER

These tanks come in a variety and can be modified MC-306s, 307s and dry bulk containers such as the one shown in **Figure 2-70.** They may carry the hot material with a mechanism to keep the material hot or be loaded hot and then heated prior to offloading. Common products are tar, asphalt, molten sulfur, and fuel oil #7 and #8.

SAFETY The major problem with these tankers is the heated material itself: Anyone coming into contact can be seriously burned. The molten material could ignite the truck or other combustibles.

If the material is allowed to cool, it can cause problems for the responders or the shipping company. Tar trucks may be transported with a propane or fuel oil flame ignited to heat the product en route to the job site, although this may be an illegal act in most states.

Manway

Discharge Points

Figure 2-69 Dry bulk tanks carry a variety of products; some examples include fertilizers, explosives, and concrete.

Figure 2-70 These trucks carry molten products and can be heating the product while driving. This practice is illegal but is found on occasion. The fuels used to heat the product are either diesel/kerosene or propane.

Intermodal Tanks

These tanks are increasing in use and carry the same types of products as their highway and rail companions. They are called intermodal as they can be used on ships, rail, or highway like the example in **Figure 2-71.**

> **CAUTION** Smaller intermodal tanks may be found inside of box-type tractor-trailers.

Intermodals follow three basic types: non-pressure, pressure, and high pressure. They are built in two fashions, either seated inside a steel frame,

Figure 2-71 This is an IMO101 tank. Like the totes, these are bulk tanks capable of carrying a large quantity of product. These are normally placed on ships, then delivered locally by a truck, although trains can also be used.

called a box-type framework, or with the tank as part of the framework, called a beam-type intermodal. Like tank trucks, they are assigned specification numbers, IM (Intermodal) 101, IM 102, Spec 51 (specification 51). When used internationally, they are called IMOs. They are made to be dropped off at a facility and, when empty, picked up for refilling. Examples of intermodal containers are provided in **Figure 2-72, Figure 2-73, Figure 2-74, Figure 2-75, Figure 2-76, Figure 2-77, Figure 2-78, Figure 2-79,** and **Figure 2-80. Table 2-7** describes IMO containers.

The piping for the IM 102 is contained outside of the tank, usually on the ends, with a manhole on top of the tank. The piping for the IM 101 is contained with a housing on top of the tank. For all IMs, the tank and valves are contained within the structure to avoid damage during transport.

Rail Transportation

As with highway transportation there are only a few types of railcars. The piping and shape of railcars may resemble their highway counterparts but that is the extent of the similarities.

> **NOTE** In rail transportation the quantities are greatly increased—up to 30,000 gallons for hazardous materials and 45,000 gallons of nonhazardous materials.

Rail incidents usually involve multiple railcars, while highway incidents usually involve one or two trucks. The incidents may occur in rural areas, away from water supplies and easy access. Rail incidents will involve multiple agencies, and the

Figure 2-72

Figure 2-73

Figure 2-74

Figure 2-75

Figure 2-76

Figure 2-77

Figure 2-72 to Figure 2-77 Shown in these photos are all of the types of intermodal tanks that are carried predominately on ships, then moved by highway or rail to their final destination. Similar to highway tanks, they carry a variety of products and will be placarded and labeled if carrying hazardous materials. *(Photos by Dan Law)*

Figure 2-78

Figure 2-79

Figure 2-80

Figure 2-78 to Figure 2-80 Shown in these photos are all of the types of intermodal tanks that are carried predominately on ships, then moved by highway or rail to their final destination. Similar to highway tanks, they carry a variety of products and will be placarded and labeled if carrying hazardous materials. *(Photos by Dan Law)*

TABLE 2-7

IMO Containers			
Container	**Maximum Capacity**	**Pressures**	**Products**
IM 101 or IMO Type 1	6,300 gallons	25.4 to 100 psig	Nonflammable liquids, mild corrosives, foods, and other products
IM 102 or IMO 2	6,300 gallons	14.5 to 25.4 psig	Flammable materials, corrosives, and other industrial materials
Spec 51 or IMO type 5	5,500 gallons	100 to 500 psig	Liquefied gases, much like an MC-331
Specialized tanks	Varies	Varies	Includes cryogenic tanks and tube banks (small tube trailer) that carry the same products as their highway counterparts.

RIGHTING OF OVERTURNED TRUCKS

The righting of overturned tank trucks can be controversial at times. The general rule is that anytime a truck is overturned, it must be offloaded prior to the truck being righted. There is substantial risk of tank failure if a loaded truck is righted by a tow truck. Some tow companies use air bags and have been trained to right fully loaded tank trucks, but it is not advisable to allow them to do so. The towing company may be aggressive in wanting to right the truck, but the incident commander is in charge, and if the tank were to fail and the product entered a waterway, the incident commander could be held responsible for violations of the Clean Water Act.

A tank truck is designed to function with its wheels down and when it is wheels up, the strength of the tank is in question. The stresses on a tank during a rollover cannot be seen and when rolled over there are more stresses placed on the tank. Gasoline tankers (MC306/DOT 406) are especially susceptible to failure during a righting operation.

The principle responsible party is required by law to provide response experts, either from within their company or an outside contractor to mitigate this type of incident. In some cases the local public sector HAZMAT team would remain in command and monitor the scene until the scene is no longer an emergency. In some parts of the country the local HAZMAT team is aggrressive in this mitigation, and performs the offload. No matter who handles the incident the product needs to be offloaded from the truck. There are several methods of removing the product from the tank, some easier than others. One method is to remove the product through existing valves and piping, which can sometimes be awkward when the truck is upside down. Another method is to attach a valve assembly to the existing valves, and then offload the product. Comparable to this method are flexible bladders which attach to the valves and are connected to piping. Another method is to drill the tank to allow for the offload of the product. Each of the methods have benefits and pitfalls, and may not apply in every case. First responders can speed up this process by calling for another compatible tank truck to transfer product into and calling for a vacuum truck or pump to conduct the transfer. These requests should be made in consultation with the responsible party and the HAZMAT team. In some states box trailers that have any quantity of chemicals, including household commodities, are required to offload prior to righting, as shown in **Figure 2-81.**

local community can expect a large contingent of assistance coming from the state and federal government, which in itself can be difficult to manage.

Railcars come in three basic types: non-pressure (general service), pressure, and specialized cars. Although they are categorized in this fashion, the commodities they carry will determine the ultimate use of the car. As with highway transportation, dedicated railcars are marked with the products they carry.

CAUTION The term **non-pressure car** is actually misapplied as it can have up to 100 psi in the tank.

It carries chemicals, combustible and flammable liquids, corrosives, and slurries.

NOTE The easiest way to determine if a railcar is non-pressure is if the valves, piping, and other appliances are located outside of a dome cover (see **Figure 2-82**).

In some cases, the car will be bottom unloaded, with the ability to load the car through the top. It is possible that the car will be top loaded and unloaded. The cars will usually have a small dome cover, but the relief valves and other piping is located outside of this dome. Most of the piping is located on the catwalk on top of the car. Corrosive cars, such as the one shown in **Figure 2-83,** will be smaller than other non-pressure cars and may have a characteristic strip of protective coating around the middle of the car.

Non-pressure cars can be insulated and may have heating and cooling lines around the tank. Prior to offloading the tank may need to be heated to ease the offloading process. There exists a non-pressure railcar that is known for its expansion dome. The piping valves and fittings will be sitting on top of this dome, which was constructed to hold any potential expansion. These cars are not in regular service but may be seen in larger industrial facilities that have their own railroad service onsite.

Pressure tank cars as shown in **Figure 2-84** also carry a wide variety of products, which include flammable gases like propane and poisonous gases like chlorine and sulfur dioxide. The pressure cars will have pressures in excess of 100 psi up to

(A)

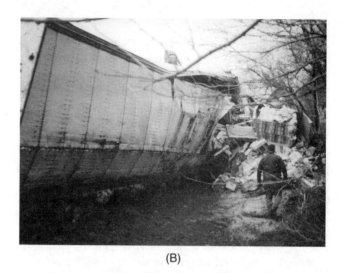

(B)

(C)

Figure 2-81 From this view (A), it would appear easy to put this truck back on its wheels for removal. The view provided by (B) shows that the roof and sides of the truck are severely damaged and will not hold back the cargo. The top and sides of trailers are merely a thin skin and offer no structural integrity. The wreck and resulting shift in cargo caused the roof and side to fail. If there are any hazardous materials in the cargo, their release would move quickly into the storm drain (C). Any discharge into this drain is a violation of the Clean Water Act. *(Courtesy of Maryland Department of the Environment)*

Figure 2-82 The indication that this is a non-pressure railcar is the fact that all of the valves are on the outside of the tank and not contained within any protective housing.

Figure 2-83 The striping around the tank indicates that the car carries corrosive materials. The coating protects the tank from spillage.

NOTE A pressure railcar has all of the valves, piping, and fittings located under the dome on top of the railcar.

---

600 psi. Most pressure cars that carry flammable materials will be insulated with a spray-on insulation or may be thermally insulated. This insulation is 1–2 inches thick and helps reduce the chance of a BLEVE during a fire situation.

A catwalk around the dome provides for relatively easy access.

Like the highway tank truck system railcars are also listed by specification number. These numbers

Figure 2-84 The pressure car is indicated by the valves being contained within the protective housing of the rail car and no exposed piping, or valving at the bottom of the rail car.

are stenciled on the side of the car and indicate some characteristics of the car. See **Table 2-8.**

SPECIALIZED RAILCARS

Specialized railcars have the same characteristics as highway transports; in fact, some highway box trailers are loaded onto flatbed railcars. When transported in this fashion they are referred to as trailers on flatcars (TOFC). You can find regular box trailers and refrigerated trailers on flatcars. Freight boxcars haul the same products as highway trailers, with the exception of carrying much larger quantities; examples of these cars are shown in **Figure 2-85.** The boxcars can carry boxes, cylinders, bulk tanks, totes, and super sacks. Refrigerated railcars, called *reefers,* are similar to highway and intermodal boxes and contain their own fuel source. Dry bulk closed railcars are common, and open hopper cars may also be found. Tube trailers for rail are rarely used; the more common mode is a highway tube trailer set on a flatcar. Cryogenics are also carried in railcars and will be the same low pressure (25 psi) as in highway transportation, with the characteristic venting when the pressure increases.

MARKINGS ON RAILCARS

Railroads use the same placarding system as highway, with the exceptions noted in the placard section. The placards are the same size, but additional information on the railcars is marked better than their counterparts on the highway. The information on the car itself is printed larger as compared to highway transport. In addition, by looking at the sides of a railcar, you can tell the specification type, maximum quantities, test pressures, relief valve settings, and other pertinent information. In addition to a placard the name of the hazardous materials will be stenciled on two sides of a **dedicated railcar.** These types of markings are shown in **Figure 2-86.**

TABLE 2-8

Railcar Specifications	
Specification number	**Type of car**
DOT105	Pressure
DOT109	Pressure
DOT112	Pressure
DOT114	Pressure
DOT120	Pressure
DOT103	Non-pressure
DOT104	Non-pressure
DOT111	Non-pressure
DOT115	Non-pressure
AAR201	Non-pressure
AAR203	Non-pressure
AAR206	Non-pressure
AAR211	Non-pressure
DOT113	Cryogenic
AAR204	Cryogenic
AAR204XT	Cryogenic
DOT106	Miscellaneous
DOT107	Miscellaneous
DOT110	Miscellaneous
AAR207	Miscellaneous
AAR208	Miscellaneous

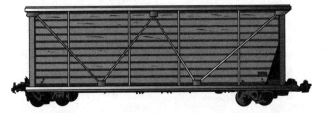

Figure 2-85 The specialized railcars are similar to their highway counterparts but transport greater quantities.

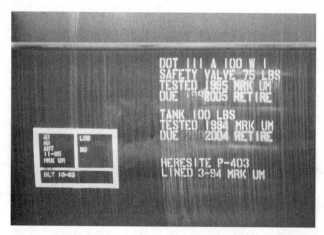

Figure 2-86 A railcar has a lot of information stenciled on the side of the car.

> **NOTE** Certain railcars may be painted in a configuration to identify their hazardous loads.

It was once common to paint a car carrying hydrogen cyanide white with a red stripe running around the middle; it was called a "candy-striper."

The specification information listed on the car is coded to reveal a lot of information about the car. The tank's owner can be determined by the code on the top line, which consists of four letters followed by a series of numbers. If the fourth letter is an *X,* then the car is owned by someone other than the railroad. The numbers indicate the car number and can be cross-referenced to determine the tank's contents when contacting the railroad. An example of the coding on a railcar is as follows:

DOT 111 A 60 AL W 1
The DOT is the authorizing agency.
111 is the class or specification number.
A has information about the rail car (applies to 105, 112, 113, 114, and 111).

- ■ A—has top and bottom shelf couplers.
- ■ S—has features of an A car and head puncture resistance.
- ■ J—has features of A and S, and jacketed thermal protection.
- ■ T—has features of A, S, and J plus spray-on insulation.

60 is the pressure of the tank.
AL is the material that the tank is made up of:

- ■ No letter—carbon steel
- ■ AL—aluminum alloy
- ■ A—carbon steel
- ■ A-AL—aluminum alloy
- ■ AN—nickel

- ■ B—carbon steel, elastomer lined
- ■ C, D, or E—alloy steel

W is the construction type of the tank.

- ■ W—fusion welded
- ■ F—forged welded
- ■ X—longitudinally welded

1 is a number that describes fittings, materials, and linings.

Bulk Storage Tanks

Bulk storage tanks come is sizes from 250 gallons up to the millions of gallons; they store a variety of products, the most common being petroleum products. These tanks are seen in residential homes, rural areas, and industrial facilities, such as the example in **Figure 2-87.** In residential homes the most common tank is a 250-gallon home heating oil tank, and some homes or small businesses may have gasoline or diesel fuel tanks that vary from 250 to 500 gallons. There are two basic groupings of tanks: in-ground and aboveground. Since the passage of the EPA underground storage regulations, there has been considerable movement to remove **underground storage tanks (UST)** and replace them with **aboveground storage tanks (AST).** If the facility owner elects to keep tanks underground, then there are additional requirements placed upon them for testing, containment, and leak prevention.

> **NOTE** Leaks from a UST can be overwhelming as the leak may not be detected for a period of time and will spread throughout the area, contaminating a large area.

Figure 2-87 The sizes of tanks vary from a few hundred gallons up to several million gallons. A catastrophic failure of the tank may overwhelm the responders' ability to handle the incident.

A substantial release from a million-gallon tank of gasoline would require an enormous response from the emergency responders and environmental contractors.

Underground tanks are usually constructed of steel and are coated with an anticorrosion material. New tanks are usually double-walled, a tank within a tank to prevent any spillage into the environment. The piping comes up through the ground, and its use will determine its route. A UST at a gasoline station will have the offloading piping come up through the pump, as **Figure 2-88** demonstrates. The piping is manufactured so that in the event that a car knocks off the pump it will shut off the flow of gasoline and will snap the piping off at or near the ground level. The only spillage of gasoline would be from the amount in the piping aboveground and in the hose, if everything works properly; see the accompanying case study for more information. The loading piping is located, separately and all of the fill pipes are located in the same area for easy transfer from a tank truck. The fill pipes are usually color-coded and marked so that the driver can differentiate between unleaded, unleaded super, and diesel fuel, although the color coding is not standard across the country and varies company to company. At some location on the property will be vent pipes for the tanks, generally away from the pumps and near the property line. Many facilities are being retrofitted with vapor recovery systems so there will

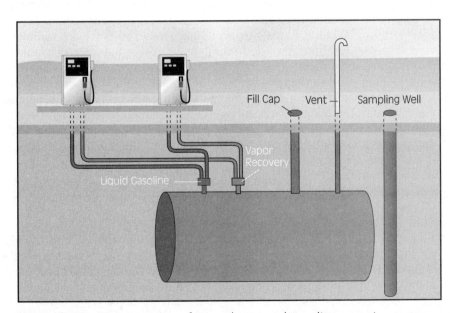

Figure 2-88 Piping system of an underground gasoline pumping system.

not be any vapors released to the air. The vapors travel back into the tank through vapor recovery lines. There will also be other manhole covers approximately 6 to 8 inches in diameter with a triangle on the top of the cover. These are inspection wells and typically surround the tank. The holes are drilled at various depths so that leaks can be easily detected by air monitoring of the well or by taking a sample if water is in the well. If a facility has had a leak, there may be a large number of these wells on and around the property. Most gas station tanks hold 10,000 to 25,000 gallons; if they are not properly monitored, they can slowly release a substantial amount of product over a short period of time.

NOTE It is not uncommon to find gasoline bubbling up in a basement miles away from a gas station from a release several years ago.

Another common problem occurs when farms are redeveloped to become housing developments; if there were unknown USTs on the property that were not discovered during construction, these tanks may eventually leak and cause serious problems. The tank and environmental industry refers to a **leaking underground storage tank** as a **LUST.**

Aboveground tanks have been used for many years but are becoming more and more common; an example is provided in **Figure 2-89.** They are of two methods of construction: upright and horizontal. ASTs can be non-pressure (atmospheric tanks) or pressure tanks. ASTs hold a wider variety of chemicals than their UST counterparts, but petroleum products are still a leading commodity stored in these tanks. They vary in size from the 250-gallon home heating oil tank to oil tanks of several million gallons. Industrial and commercial applications are required to have a containment area around the tank. The containment area must

CASE STUDY

A recent incident in Baltimore County, Maryland, demonstrates that no matter how many protection systems are in place, a chemical release off the property is still possible. On an early afternoon a gas station was getting a tank filled with fuel; the tank filling went as expected with no problems. The gas station had installed a state-of-the-art alarm system that would indicate a fault in the system and alert to any potential releases. The system goes through a system check to ensure that the system is intact and there are no releases. This system check is completed after each sale, if no pump is operated for a period of 30 seconds. If a pump is on or the period between sales is less than 30 seconds, then the system check does not begin to work. During a busy afternoon, it could be estimated that the system check would not function for a long period of time, as at least one of the pumps is operating all the time. In addition to the system check, each of the storage tank locations, some piping, and the pumps all have electronic monitors that detect the presence of a liquid. When a liquid is detected, an alarm activates, the same type of alarm as if the system check fails the system. One of the potential problems with many self-service gas stations is that there is only one attendant, who for many reasons cannot leave the work area to check on a problem outside or, depending on the alarm panel location, may not be able to check any alarms. In some gas stations the alarms are in other rooms and cannot

be seen or heard by the only attendant. Another problem is that the liquid alarms will activate when it rains, tripped by rainwater.

The pumps that sit on top of the storage tanks can supply upwards of 80 psi of pressure to all pump station nozzles, although this is governed down to an actual working pressure of 10 psi at all pumps. In this particular incident it appears that a flange that comes out of the tank pump failed and separated from the tank pump. From the pump discharge there is a 2-inch diameter opening, which at 10 to 80 psi could pump a considerable amount of fuel in a short time. One theory is that a customer may have set a nozzle down after not being able to pump fuel or the amount of fuel being pumped may have been inadequate. This action would have resulted in the tank pump continuing to pump. A liquid alarm sounded in a number of areas at the gas station. The attendant notified the pump repair company, per protocol, but a response was delayed due to a number of extenuating circumstances. Once the repair company arrived, the crew found the tank empty, and the clerk informed them of the recent delivery. It was later found that an estimated 4,500 gallons was lost, causing flammable vapor readings in a several block area, although all the gasoline was later recovered. Within the gas delivery system, this station had the best protection in place and had just retrofitted the station with a new piping system. No matter how many protection systems are in place, it is still possible to have a release.

Figure 2-89 Due to increased environmental concerns many tanks are being placed above ground and are called aboveground storage tanks, or AST.

Figure 2-91 HAZMAT crews were on location where a ship rammed an offshore transfer pipe location. From across the river they saw the roof of this tank fly up into the air, heard the explosion, and saw the tank erupt in flames. Upon arrival it was critical to determine the contents of the tank and adjacent tanks. The tier 2 chemical inventory would have this information. Complicating the extinguishment of this asphalt tank was that it is an insulated tank, with fire in between the outer skin and the tank walls. *(Photo by the Maryland Department of the Environment)*

be able to hold the contents of the largest tank within the containment. Regular inspection of these areas is required to ensure that rainwater, snow, or ice does not cause a buildup of liquid in the containment area which would reduce the containment's ability to hold its intended amount. Depending on the weather conditions the containment area's gate valve may be left in an open position, which leaves the facility at risk for product to escape the facility through the open drain or gate valve.

Upright storage tanks come in three basic construction types: **ordinary tank, external floating roof tank,** and **internal floating roof tank.** The ordinary tank (see the example in **Figure 2-90**) is constructed of steel and typically has a sloped or cone roof to shed rainwater, snow, and ice. The roof and tank shell interface are purposely made weak so that in the event of

an explosion, only the top of the tank is relocated, as is shown in **Figure 2-91.** One of the major problems with this type of tank is when the tank is not full, there is room for vapors to accumulate.

CAUTION Any time vapors are allowed to accumulate, there is potential for a fire or explosion.

In chemical process facilities ordinary tanks are constructed purposely without a roof in place; they are typically used as a safety outlet to dump a chemical product into a holding tank. It is not uncommon to have a storage tank with piping into the tank just to catch overflow or the contents of a system in the event of overpressurization or failure. As the product may be hot, producing considerable vapors, or reacting, it is desirable to leave this tank uncovered. Water treatment areas of chemical plants may contain chemicals that need to evaporate in the wastewater tank, and in many of these systems the tanks are left uncovered. From a planning perspective it is important to know what type of tanks exist in a facility, as many tanks may be in place for safety reasons. It is important to identify these tanks during pre-incident surveys so as not to misjudge the severity of an incident. In the event of a fire it would be a

Figure 2-90 The ordinary tank has a weak roof-to-shell seam, so in the event of an explosion, the roof will come off, but the tank should remain intact.

Figure 2-92 The roof of this tank floats on top of the liquid, which limits the production of vapors and minimizes the fire potential.

waste of resources to try to cool a tank, only to find out that the tank was not holding product and was merely a safety outlet.

External floating roof tanks are used to eliminate the buildup of vapors, as they ride on top of the liquid, such as the one shown in **Figure 2-92.** Since there is no space, vapors cannot accumulate and create problems. External floaters can be seen from the top of the tanks, and there is a ladder affixed to the roof to allow access, as shown in **Figure 2-93.**

NOTE A common incident with these types of tanks results from a lightning strike that causes a fire in the roof/shell interface area. This type of fire is difficult to extinguish and can result in the roof/tank failure if not controlled quickly.

Figure 2-93 A floating roof tank providing a view of the floating roof. As the roof moves up and down the ladder moves along. A common fire problem is a lightning strike; what results is a seal fire where the tank wall meets the roof.

It is also possible that water used for firefighting could sink the roof, causing further problems.

Internal floaters are constructed in the same manner as the external floaters but have added an additional roof over the top of the tank, as the example in **Figure 2-94** shows. The type of roof varies but is usually a slightly coned roof with vent holes along the outer edge of the tank shell or a geodesic type of roof that is usually made of fiberglass. The internal floater suffers from the same risk of fire problem as the external, although it is a reduced risk. If a fire were to start in the roof/shell interface, the roof would make it more difficult to extinguish.

SPECIALIZED TANKS

These tanks are a combination of the tanks discussed earlier in this section. They include pressure tanks and cryogenic tanks. The larger pressure vessels are divided into two categories, low pressure and high pressure tanks. The low pressure tanks hold flammable liquids, corrosive liquids, and some gases, up to 15 psig. Common high pressure commodities are liquefied propane, liquefied natural gas, or other gaseous or liquefied petroleum gases. An acid like hydrochloric acid, which has a high vapor pressure, would not be uncommon in this type of tank. These tanks may have an external cover, which appears to be a tank within a tank. The pressure tanks are larger version of the propane tanks shown earlier and can have a capacity of up to 9 million gallons, although less than 40,000 gallons is typical. Liquefied petroleum gases are not only stored in just pressure tanks, such as the one shown in **Figure 2-95,** but these types of gases may be stored in other locations, such as caves carved out of mountains, although this is rare. As these facilities are not required to report under the SARA Title III regulations, responders

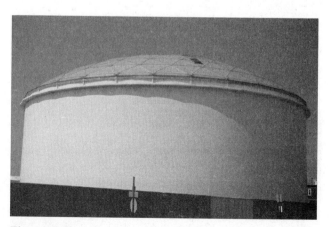

Figure 2-94 The internal floater is constructed in the same fashion as the external floating roof. It just has a cover on it to keep out rain and debris.

Figure 2-95 A specialized tank, such as this propane tank, has some of the same properties as transportation equivalents.

Figure 2-96 This tank holds refrigerated liquefied gases, in particular oxygen.

must contact local gas suppliers to find out how they store their gas products.

Upright cryogenic storage tanks are located in almost every community, especially if the community has a hospital or medical center, such as those shown in **Figure 2-96.** Most hospitals are supplied with LOX through the cryogenic tank. Facilities that sell or distribute compressed gases are likely to have cryogenic tanks. Fast-food restaurants and convenience stores are now using cryogenic tanks to supply carbon dioxide to soda dispensing machines.

SENSES

The use of vision and hearing are acceptable senses that could be used while investigating potential chemical releases. The use of touch, smell, and taste are a dangerous use of senses.

CAUTION The lack of an odor cannot be equated to a lack of toxicity, as many severely toxic materials are colorless and odorless.

Responders can use these sensory clues gained from other persons to help identify the spilled material. The smell of materials is an important clue when trying to identify a spilled material. If bystanders or evacuated persons can assist by describing an odor, take advantage of this clue, but do not endanger your life to determine an odor. Many toxic materials can be harmful if touched as the material can be absorbed through the skin. Anytime people handle a chemical, they should wear appropriate chemical protective clothing.

CHEMICAL AND PHYSICAL PROPERTIES

Although not intended to be a full chemistry lesson, this next section covers a key safety component. The chemical and physical properties outlined here are appropriate for your level of response. These identification and use of these key terms can determine the outcome of the incident—and your well-being. As you progress through the response levels, the need for additional chemistry also increases. When in doubt, consult with your HAZMAT team or other resources such as Chemtrec or a local specialist. As you will learn, many of the terms discussed here can be applied through basic firefighting. Fire is a chemical reaction; thus, the better you understand this chemical reaction, the better off you will be and the more you will understand how conditions can rapidly change, which will also benefit the citizens of your community.

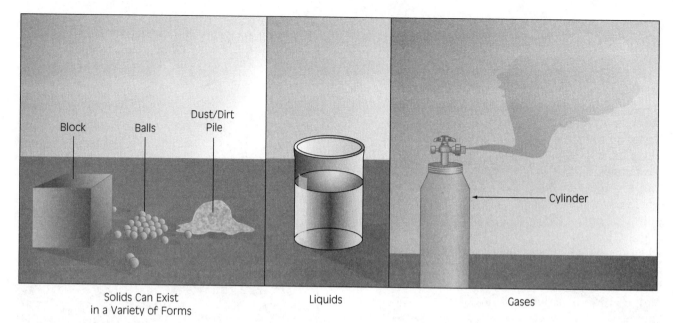

Figure 2-97 States of matter.

States of Matter

The basic chemical and physical properties that are important to understand are the **states of matter—solid, liquid,** and **gases**—as shown in **Figure 2-97.** The severity of the incident can be determined by knowing if the material is a solid chunk of material, a pool of liquid, or an invisible gas.

> **NOTE** The level of concern rises with each change of state; relatively speaking a release of a solid material is much easier to handle than a liquid release, and it is nearly impossible to control a gas.

The control methodology for each increases in difficulty from simple controls with a solid to difficult with a gas. Evacuations must be enlarged for releases involving gases, where a minimal evacuation would be required for a solid material.

The route that the chemicals can harm humans also varies with the state of material. Solids usually can only enter the body through contact or ingestion, although inhalation of dusts is possible. Liquids can be ingested, absorbed through the skin, and, if evaporating, inhaled. Gases, on the other hand, can be absorbed through the skin, inhaled, and, to some extent, ingested.

Related to states of matter are melting point, freezing point (shown in **Figure 2-98**), boiling point

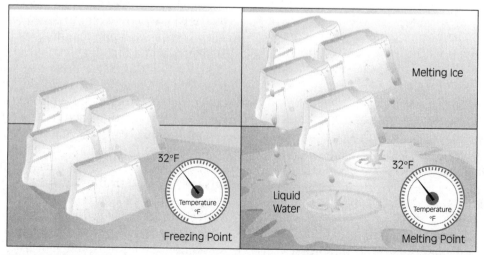

Figure 2-98 Melting and freezing points.

(shown in **Figure 2-99**), and condensation point. All of these are points at which a material changes states of matter. The ability of a material to go from a solid to a gas is called **sublimation.** Some common chemicals that sublimate are moth balls and dry ice (carbon dioxide in solid form). The **melting point** is the temperature at which a solid must be heated to transform the solid to the liquid state. Ice has a melting point of 32°F. The **freezing point** is the temperature of a liquid when it is transformed into a solid. For water the freezing point is 32°F, which may seem confusing as it is also the **melting point.** The actual temperatures vary by tenths of a degree, but are very close. The **boiling point** is when the liquid is heated to a point that evaporization takes place, and the liquid is changed into a gas. Another way of defining boiling point is the temperature of a liquid when the vapor pressure exceeds the atmospheric pressure and a gas is produced. Water boils at 212°F and changes into a gaseous state. The important thing to remember about boiling points is the fact that when the liquid approaches this temperature, vapors are being produced, which can cause serious problems.

Vapor Pressure

Out of all a hazardous material's chemical and physical properties, vapor pressure can be critical to a hazardous materials responder. If a product has a high **vapor pressure,** it can be very dangerous to deal with. A material with a low vapor pressure is typically not a major concern. Vapor pressure has to do with the amount of vapors that are released from a liquid or a solid, as shown in **Figure 2-100.** The true definition is the pressure that is exerted on a closed container by the vapors coming from the liquid or solid. You can relate vapor pressure to the ability of a material to evaporate, as the material is not disappearing but moving to another state of matter. Chemicals like gasoline, acetone, and alcohol all have high vapor pressures, while diesel fuel, motor oil, and water all have low vapor pressures. Vapor pressures are measured in three ways: millimeters of mercury (mm Hg), pounds per square inch (psi), and atmospheres (ATM). Normal atmospheric conditions are listed here; materials with vapor pressures at or above these levels are usually gases.

- 760 mm Hg
- 14.7 psi
- 1 ATM

Although these figures are used to describe normal vapor pressure, they are best described as atmospheric pressure. The standard temperature is 68°F (20° C). If a temperature is not provided, then it can be assumed to be 68°F. Temperature has a direct relationship with vapor pressure. The higher the temperature, the more vapors will be produced. Conversely, the lower the temperature, the lesser amount of vapors that will be produced. In situations such as a chemical spill in a building, the decrease in temperature can aid first responders by reducing the amount

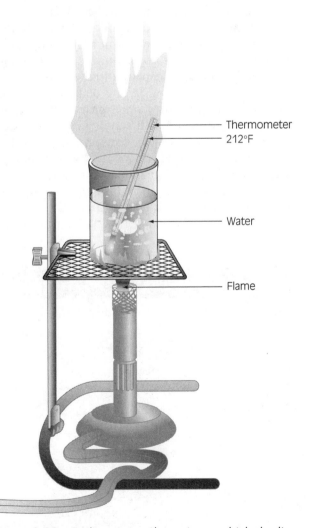

Figure 2-99 Boiling point, the point at which the liquid is turning into a gas.

Figure 2-100 A material with a vapor pressure is pushing against the sides of the container. If the vapor pressure is high and the material is in the wrong type of container, the container could fail.

of vapors. HAZMAT responders can use elevated temperatures to remove a chemical from a building, and this tactic is used occasionally. Chemicals that have a significant vapor hazard are those in excess of 40 mm Hg; they are considered volatile. Chemicals with a vapor pressure of less than 40 mm Hg do not present much inhalation hazard even though they can still present extreme toxicity through skin absorption. Chemicals with a vapor pressure above 40 mm Hg can be considered inhalation hazards in addition to any other route of exposure they may possess. **Table 2-9** lists the vapor pressure and boiling point of several chemical compounds.

We can use a comparison to water to judge a material's vapor pressure, as we know how quickly water evaporates. This comparison is done at 68°F, and elevated or depressed temperatures play a factor in how quickly a material evaporates. In general materials with a vapor pressure less than water do not present a significant vapor risk as the amount of vapors being produced is small. The amount of product spilled and the location of the spill must be considered. A cup of water will take at least a couple days to evaporate. The same amount of gasoline will take a few hours to evaporate, which means that vapors (therefore an inhalation risk) are being produced at a faster rate. Ethyl ether will only remain in the liquid form for a few minutes as it has a high vapor pressure. Materials with vapor pressures such as liquid chlorine, which must be put under pressure to liquefy, will instantly turn to a gas upon release. On the other hand, sulfuric acid has a very low vapor pressure and really only presents a contact hazard. It would take a long time for the acid to turn to the vapor form. Sodium hydroxide needs to be heated to 2534°F for even a small amount of vapor to be released. In reality these two materials present a minor vapor hazard, as they have an extremely low vapor pressure. We know as responders that when something is heated to 2534°F, it is probably on fire and there are other issues to contend with other than a very small amount of sodium hydroxide vapors. The pesticide ethion has an even lower vapor pressure and will remain in liquid form for a long period of time, extending to months. The true hazard from ethion is from contact, not from inhalation. Vapor pressure along with the use of air monitors is the key to risk-based response, detailed in Chapter 5.

Vapor Density

It is easy to confuse vapor pressure with vapor density, but they have entirely different meanings. **Vapor density** determines whether the vapors will rise or fall, as shown in **Figure 2-101**. When deciding on potential evacuations and other tactical objectives (such as sampling and monitoring), this is an important consideration. One of the primary reasons that natural gas leaks do not ignite or flash back more often is the vapor density of natural gas, which has a tendency to rise up and away. Air is given a value of 1, and we compare all other gases to air. Gases that have a vapor density of less than 1 will rise in air and dissipate, while gases with a vapor density greater than 1 will stay low to the ground. Natural gas has a vapor density of 0.5, while propane has a vapor density of 1.56, which causes it to seek out low spots like gullies or sewers. Propane is more likely to ignite or flash back as it has a greater potential of finding an ignition source.

TABLE 2-9

Common Products and Their Vapor Pressures		
Name	**Vapor pressure at 68°F**	**Boiling point**
Water	25 mm Hg	212°F
Acetone	180 mm Hg	134°F
Gasoline	300–400 mm Hg	399°F
Ethyl ether	440 mm Hg	94°F
Chlorine	6.8 ATM (5168 mm Hg)	−29°F
Diesel fuel	5 mm Hg	304–574°F
Methyl alcohol	100 mm Hg	149°F
Sodium hydroxide	1 mm Hg @ 2534°F	2534°F
Sulfuric acid	0.001 mm Hg	554°F
Ethion (pesticide)	0.0000015 mm Hg	304°F (decomposes)
Sarin nerve agent	2.1 mm Hg	316°F

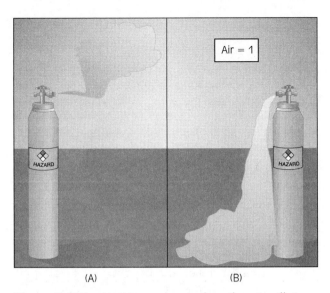

Figure 2-101 (A) Vapor density less than 1 will rise in air. (B) Vapor density greater than 1 will sink.

Specific Gravity

This chemical property is similar to vapor density, as it determines whether the material sinks or floats in water, as shown in **Figure 2-102. Specific gravity** is of prime concern when efforts are being taken to limit the spread of the spill by the use of booms or absorbent material. Water is given a value of 1, and chemicals that have a specific gravity of less than 1 will float on water. Fuels, oils, and other common hydrocarbons (chemicals comprised of hydrogen and carbon, typically combustible and flammable liquids) will float on water and have a specific gravity of less than 1. Other materials that have a specific gravity of greater than 1 will sink in water and will make their way to the bottom. It is much easier to recover materials floating on top of the water than those underwater. Carbon disulfide

and 1,1,1 trichloroethane are two common materials that sink in water. Any material that sinks in water is especially dangerous when it has the potential of reaching any groundwater, as remediation efforts are difficult and expensive. Also of concern is a material that is water soluble, or has the ability to mix with water, not sink or float. Corrosives and many poisons are water soluble and will be difficult to remove from a water source; two examples are provided in the accompanying text box. Alcohol is also water soluble, and as you will learn with fire-fighting foam, materials that are water soluble are difficult to extinguish.

Corrosivity

We have covered at length tanks and containers that hold corrosives, so you should already be aware of some of the concerns in dealing with corrosives. This is a term that is applied to both acids and bases and is used to describe a material that has the potential to corrode or eat away skin or metal. We deal with corrosives every day: Our body is naturally acidic, and we use many corrosive materials in our everyday lives. Acids are sometimes referred to as corrosives, while bases are also known as alkalis or caustics. The accurate way to describe a corrosive is to identify the material's pH, which provides some measure of corrosiveness. Neutral materials are given a pH of 7, while acids range from 0 to 7, and bases from 7 to 14. Although in the lab it may be important to differentiate between a pH of 1.1 and 1.7, the resulting red mark from being splashed with the material will be the same. The issue of concentration is an important one when dealing with corrosives as it affects how that particular corrosive may cause harm. As an example, Pepsi™ has a pH of 2 but is not hazardous because the corrosive phosphoric

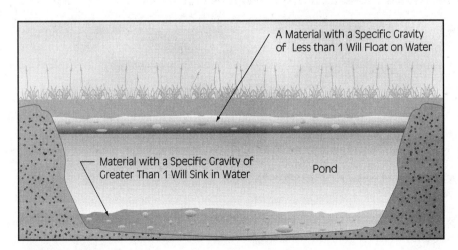

Figure 2-102 Specific gravity.

CASE STUDY

When dealing with water-soluble materials, it is important to protect bodies of water. Cleanup can be difficult, even impossible, if the material enters the water. In one incident more than 10,000 gallons of sodium hydroxide, a very corrosive material, entered a small stream. The creek had to be dammed and treated with a neutralizing agent. Any life existing in this mile of stream was killed by the sodium hydroxide, but the environmental damage could have spread for many more miles if it had not been neutralized. Once neutralized, the water was released and eventually led to a larger body of water, where no damage occurred. This took considerable resources and time to accomplish; luckily, the necessary neutralizing agent was located at the site of the spill.

In another incident a very small amount of a pesticide was sprayed near a pond. After a rainfall the pesticide ran into the pond, killing all of the fish in the pond. The only method of removing this pesticide is the addition of other chemicals, which are also hazardous. It will be years before the pond can sustain life.

In Baltimore, Maryland, a railcar was involved in an accident that released 18,000 gallons of hydrochloric acid, which eventually ran into a local stream. The addition of the acid actually brought the pH level of the stream to near acceptable levels, as this stream was already heavily contaminated from a number of other sources. Cleanup of these three incidents required the express permission of local or state environmental agencies.

acid is not concentrated. The bottom line is that if the phosphoric acid were concentrated, it would have more potential to cause harm. The impact upon humans by a corrosive material is best evaluated by a HAZMAT technician. The first responder should avoid any potential contact with corrosive materials in an emergency response setting. The emergency response community is understandably concerned about the potential harm posed by corrosives. **Table 2-10** lists the pH of some common compounds.

Chemical Reactivity

When mixed, chemicals will have one of three types of reactions: exothermic, endothermic, or no reaction.

NOTE The most common chemical reaction is **exothermic,** meaning the release of heat.

For example, fire is a rapid oxidation reaction (exothermic) accompanied by heat and light. When most chemicals mix and provide an exothermic reaction, there usually is not a lot of light but substantial heat may be produced. Mixing one tablespoon each vinegar (acetic acid) and ammonia (base) at the same temperature will produce an immediate rise in temperature of 10°F. When handling oleum (concentrated sulfuric acid) and applying water to a spill, the resulting mixture will bubble, fume, boil, and heat to over 300°F the instant the water hits the acid. An **endothermic** reaction is one in which the heat gener-

TABLE 2-10

pH of Common Materials			
Material	**pH**	**Material**	**pH**
Water[1]	7	Sulfuric acid	1
Stomach acid	2	Gasoline	7
Orange Juice[2]	2	Hydrochloric acid	0
Drain cleaner[3]	14	Pepsi™[4]	2
Potassium hydroxide	14	Household ammonia[5]	12

[1]Tap water is usually a 7, while rainwater in the Northeast can be 3–6.
[2]Citric acid is the main ingredient.
[3]Main ingredient is sodium hydroxide (lye).
[4]Phosphoric acid is the main ingredient.
[5]Houshold ammonia is a 5% solution of ammonium hydroxide and water.

TABLE 2-11

Flash Point of Some Common Materials			
Material	**Flash point**	**Material**	**Flash point**
Gasoline	−45°F	Diesel Fuel	>100°F
Isopropyl alcohol	53°F	Motor oil	300–450°F
Acetone	−4°F	Xylene	90°F

ated by the reaction is absorbed, and no heat is released. A cold pack is a common example.

Flash Point

Flash point is described as the temperature of the liquid that, when an ignition source is introduced into the vapor/air mixture above the liquid during heating, will produce a flash fire. This resulting flash fire will ignite just the vapors and self-extinguish once those vapors are burned up. The liquid itself does not burn; only the mixture of vapors and air is ignited. Following closely behind flash point is the fire point of a liquid. The fire point is the temperature of a liquid, which produces vapors when heated, that will ignite the liquid and sustain burning. In the lab scientists can replicate flash points and fire points; on the street, however, they are usually one and the same. **Table 2-11** lists the flash point of some common chemicals.

Auto-Ignition Temperature

Also sometimes referred to as **ignition temperature,** this is the temperature that the material will ignite on its own without an ignition source. Ignition temperatures are much higher than flash points. Once the ignition temperature is reached, the material will burn. **Self-accelerating decomposition temperature (SADT)** is another term essentially the same as auto-ignition temperature. Whatever term is used, it is a level to avoid.

Flammable Range

Two levels define **flammable range,** lower explosive limit and upper explosive limit. The flammable range is the difference between the two extremes. To have a fire or explosion you need an ignition source, a fuel (vapors), and air. The proper vapor-to-air ratio must exist to produce ignition. The **lower explosive (flammable) limit (LEL)** is the lowest amount of vapor mixed with air that can provide the proper mixture for a fire/explosion. The air monitor

used to detect combustible gases is designed to read this level. The **upper explosive (flammable) limit (UEL)** is the highest amount of vapor mixed with air that will sustain a fire or explosion. Each year there are any number of natural gas explosions throughout the United States, and when we examine the mixture required we find that the LEL of natural gas is 5 percent. In other words, when natural gas vapors are mixed with 95 percent air with an ignition source present, an explosion or fire is possible, as shown in **Figure 2-103.** The upper explosive limit for natural gas is 15 percent, which combines with 85 percent air to have a fire or explosion. The flammable range for natural gas is 5 to 15 percent, and a fire or explosion can occur at any point in that range, as long as there is enough air to complete the mixture. A combination of 4 percent methane and 96 percent air could not sustain a fire or explosion; nor would a mix of 16 percent natural gas and 84 percent air. Acetylene, which is a common gas, has a LEL of 2.5 percent and UEL of 100 percent, which means that when you reach 2.5 percent you have the potential for a fire. Less than the LEL is called *too lean,* and amounts in excess of the UEL are called *too rich.* Finding levels less than the LEL are much safer than finding levels above the UEL.

SAFETY When confronted with a situation that has a level higher than the UEL, whenever the area is ventilated you must pass through the flammable range, an extremely dangerous situation.

Ventilation should be carried on using non–spark-inducing devices, and great care should be taken to minimize any potential electrical arcs, such as using light switches or doorbells.

Toxic Products of Combustion

This is the one area in which responders suffer considerable chemical exposures. Any time you are in smoke or breath smoke, your body is being

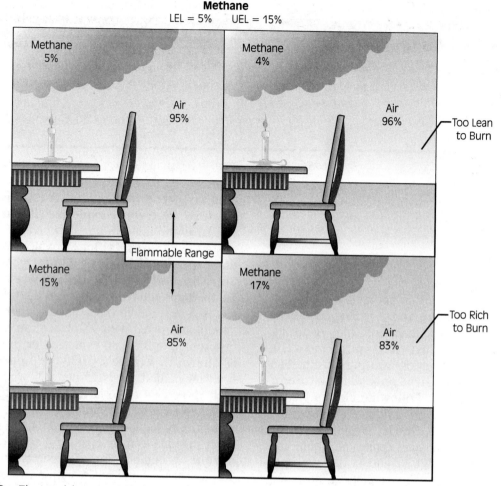

Methane
LEL = 5%   UEL = 15%

Figure 2-103 Flammable range. To have a fire or explosion the lower explosive limit must be reached. Each gas has a flammable range in which there can be a fire or explosion. Below the LEL or above the UEL, there cannot be a fire.

bombarded with toxic chemicals. A basic lesson for responders is that fire produces many toxic chemicals, such as carbon monoxide, carbon dioxide, hydrogen cyanide, hydrochloric acid, phosgene, and many others. The worst chemical accident a responder can respond to is a house, car, or dumpster fire. Even brush fires are not exempt from the toxic products of combustion, as the field may have been sprayed with pesticides, herbicides, or other chemicals. Even burning wool or hay produces extremely toxic gases. Due to this constant exposure to these and other materials, it is important to wear all of your protective clothing, especially your SCBA.

SUMMARY

The ability to recognize and identify the potential for hazardous materials is important for the first responder. The numbers of tank trucks, tank cars,

and containers can easily overwhelm the beginning student. At every incident responders must complete the recognition and identification process, whether it be the location, placards, containers, or physical senses. It could even be your sixth sense that alerts you to a potential problem; when that occurs, proceed with caution. An important lesson for responders is that you don't need to commit everything to memory, but you must know where you can access that information. It is not expected that each community have a railcar expert in its response organization. What is expected is that each agency have a contact person available around the clock to obtain that resource.

It is possible that every material on this earth has the ability to cause harm in some fashion, but chemical and physical properties plays a factor in the potential for harm. Materials that do not have a vapor pressure present little risk to responders unless touched or eaten. Materials with high vapor pressures present great risk to responders and to the community and should be treated with caution. Vapor pressure is only one of the terms that you as a responder should understand, as a variety of chemical properties may come into play. Your local HAZ-MAT responders are a good resource and should be contacted early in an incident and whenever you need assistance.

KEY TERMS

Aboveground storage tanks (AST)   Tanks that are stored above the ground in a horizontal or vertical position. Smaller quantities of fuels are often stored in this fashion.

BLEVE   Boiling liquid expanding vapor explosion, which occurs when a pressurized tank is heated and the pressure inside the tank exceeds the capacity of the relief valves to handle the pressure. When pressurized tanks rupture, they can travel significant distances, and if holding flammable materials, they will usually ignite, resulting in a large fireball.

Boiling point   Temperature to which liquids must be heated to turn to a gas.

Building Officials Conference Association (BOCA)   A group that establishes minimum building and fire safety standards.

Bulk tank   Large transportable tank comparable to but larger than a tote.

Chemtrec   The Chemical Transportation Emergency Center, which provides technical assistance and guidance in the event of a chemical emergency. This network of chemical manufacturers provides emergency information and response teams if necessary.

Cryogenic gas   Any gas that exists as a liquid at a very cold temperature, always below $-150°F$.

Deflagrates   Rapid burning, which can be considered a slow explosion, but travels at a lesser speed than a detonation.

Department of Transportation (DOT)   Federal agency that issues regulations regarding the transportation of hazardous materials.

Endothermic reaction   Chemical reaction in which heat is absorbed and the resulting mixture is cold.

Exothermic reaction   Chemical reaction that releases heat, as in when two chemicals are mixed and the resulting mixture is hot.

External floating roof tank   Tank in which the roof is exposed to the outside and covers the liquid within the tank. The roof floats on the top of the liquid, which does not allow for vapor build-up.

Flammable range   The correct mixture of flammable gas and air in which there can be a fire; for most gases there is a certain span (range) that will allow for a fire to ignite.

Flash point   Temperature of a liquid, which if an ignition source is present, will ignite only the vapors being produced by the liquid, creating a flash fire.

Frangible disk   A pressure-relieving device that ruptures to vent the excess pressure. Once opened, the disk remains open; it does not close after the pressure is released.

Freezing point   Temperature at which liquids become solids.

Gas   State of matter that describes the material in a form that moves freely about and is difficult to control. Steam is an example.

Ignition temperature   Temperature of a liquid that will ignite on its own without an ignition source; also called *self-accelerating decomposition temperature.*

Intermodal containers   Constructed to be transported by highway, rail, or ship. Intermodal containers exist for solids, liquids, and gases.

Internal floating roof tank   Tank with a roof that floats on the surface of the stored liquid; this tank also has a cover on its top to protect the top of the floating roof top.

Leaking Underground Storage Tank (LUST)   Term to describe this potential health and environmental hazard.

Liquid   State of matter that implies fluidity, which means a chemical can move as water would. There are varying states of being a liquid from moving very quickly to very slowly. Water is an example.

Lower explosive limit   The lower part of the flammable range, or the minimum required to have a fire or explosion.

Melting point   Temperature at which solids become liquids.

North American Emergency Response Guidebook (NAERG)   Reference book provided by the DOT to assist first responders in making decisions at a transportation-related chemical incident.

Ordinary tank   Horizontal or vertical tank that usually contains combustible or other less hazardous chemicals. Flammable materials and other hazardous chemicals may be stored in smaller quantities in these types of tanks.

Polymerization   A runaway chain reaction that can be violent if contained. Once started it cannot be stopped until the chemical reaction has run its course.

Relief valve   Device designed to vent pressure in a tank, so that the tank itself does not rupture due to an increase in pressure. In most cases these devices are spring loaded so that when the pressure decreases, the valve closes to keep the chemical inside the tank.

Reportable quantity (RQ)   A term used by the EPA and DOT to describe a quantity of chemicals that may require some type of action, such as reporting an inventory or reporting an accident involving a certain amount of chemicals.

Sea containers   Shipping boxes designed to be stacked on a ship and then placed onto a truck or railcar; also called sea boxes.

Self-accelerating decomposition temperature (SADT)   See *ignition temperature.*

Solid   State of matter that describes chemicals that may exist in chunks, blocks, chips, crystals, powders, dusts, and other types. Ice is an example.

Specific gravity   Value that determines if the liquid will sink or float in water. Water is given a value of 1, and chemicals with a specific gravity of less than 1 will float on water. Those chemicals with a specific gravity of greater than 1 will sink in water.

Specification (spec) plates   Label on trucks and tanks that lists the type of vehicle, capacity, construction, and testing information.

States of matter   Term applied to chemicals to describe the forms in which they may exist: solids, liquids, or gases.

Sublimation   Ability of a solid to go to the gas phase without being liquid.

Tote   Large tank, usually 250 to 500 gallons, constructed to be transported to a facility and dropped for use.

Underground storage tanks (UST)   Tanks, commonly for gasoline and other fuels, that are buried under the ground.

Upper explosive limit (UEL)   Upper part of the flammable range; above the UEL, fire or an explosion cannot occur as there is too much fuel and not enough oxygen.

Vapor density   Value that determines if the gas will sink or rise in air. Air is given a value of 1, and chemicals with a vapor density less than 1 will rise in air. Those with a density of greater than 1 will sink and stay low to the ground.

Vapor pressure   Amount of force pushing vapors from a liquid; the higher the force, the more vapors (gas) are being put into the air.

REVIEW QUESTIONS

1. What are the nine hazard classes as defined by DOT?
2. An explosive placard is what color?
3. What three things on a placard indicate potential hazards?
4. An MC 306/DOT 406 tank truck commonly carries what product?
5. A DOT 406 tank truck has a characteristic shape from the rear. What is it?
6. An MC 331 carries what type of gases?
7. What does the blue section of the NFPA 704 system refer to?
8. What are the locations of emergency shutoffs on an MC 331?
9. What are the four basic clues to recognition and identification?
10. A truck with dangerous placards is carrying what product(s)?
11. Which Table 1 placards are required for any quantity?
12. What does a material listed on the shipping papers as Packing group I indicate?
13. A tractor-trailer carrying class 1.1 materials is well involved in fire. What are your initial tactics regarding this fire?
14. In the NFPA 704 system, what does the use of ALK mean?
15. When propane tanks are involved in a fire, what is a potential consequence?
16. Which tank truck is commonly referred to as a chemical hauler?
17. How many tanks (pots) does an MC 312 have?

ENDNOTES

[1]The requirements for the subdivisions are presented in general, and there may be other requirements that need to be met for a material to be assigned to a subdivision. Refer to 49 CFR 170–180.

[2]The notation that a DOT-406 is a gasoline tanker is a general reference, and a DOT-406 can hold any number of commodities, not just gasoline. A DOT-406 may even carry food grade products or other chemicals.

Suggested Readings

Bevelacqua, Armando, and Richard Stilp, *Hazardous Materials Field Guide,* Delmar, a division of Thomson Learning, Albany, NY, 1998.

Lesak, David, *Hazardous Materials Strategies and Tactics,* Prentice Hall, New Jersey, 1998.

Noll, Gregory, Michael Hildebrand, and James Yvorra, *Hazardous Materials Managing the Incident,* Fire Protection Publications, Oklahoma University, 1995.

Schnepp, Rob, and Paul Gantt, *Hazardous Materials: Regulations, Response, and Site Operations,* Delmar, a division of Thomson Learning, Albany, NY, 1999.

Smeby, L. Charles (editor), *Hazardous Materials Response Handbook,* 3rd ed., National Fire Protection Association, 1997.

CHAPTER 3

INFORMATION RESOURCES

OUTLINE

STREET STORY

I had been on the job for several years. There was nothing unusual about the call when it first came in—nothing remarkable that would lead to any heightened sense of awareness. "Engine 29, check out the smell of smoke in the building at 211 Mason Street." It would be a simple food on the stove call or something to that effect.

When we arrived on the scene, the local police department had already made an initial entry to the facility (which turned out to be a restaurant) and reported a slight haze with a strange pungent odor but no visible fire. We made entry into the vacant facility and immediately noticed the haze and pungent odor, but because it was a restaurant, we thought it was simply something to do with the food preparation process. As we made our way through the facility, we began to notice a strange taste in our mouths and several of those present began to experience irritation to the eyes and throat, yet still nothing clicked. It was, after all, simply a restaurant, perhaps some burnt wiring or bad cooking, no big deal. As we continued our investigation, several of those present began to complain of tightness in the chest, headaches, and nausea. I was too feeling strange but did not want to throw in the towel, so another responder and I continued on.

Then suddenly as the haze began to clear via the doors and windows that had been opened for ventilation, we began to notice the dead insects, an abundance of dead insects and rodents throughout the facility. And now on the radio we heard the call for an ambulance as one of the responders who had been on the scene for some time began to experience more than simply mild irritations. We began to realize that the haze was in fact some type of aerosolized insecticide/pesticide and that we had now been exposed and contaminated. Now and only now did we begin to comprehend the error of our ways and implement the proper response and protocols.

As the weeks followed, I replayed the incident over and over and came to realize all of the mistakes that we had made. I understood how imperative it was to seek the proper training to ensure that I never put myself or those with me in this position again. Two of the responders were unable to return to work for a period of time (in one case several months). They had to be treated for the effects of organophosphates. Both are fine today but much has been written about the long-term effects of exposure to organophosphates, and one can only wonder what course these individuals' health could have taken had they been further exposed. And I do wonder, all too often.

Speaking only to my errors, there was no attempt to obtain the proper information from the appropriate resources or the owners. (After we realized we had no fire, we should have consulted the owners before further entry, because there was no life hazard.) Nor was subsequent contact made with other tenants in this row structure who may have been able to provide information or take basic actions to ensure the health and safety of those involved. The failure to properly research and gather information on this call could have resulted in much more severe health problems for all involved had the concentrations been higher or ventilation unsuccessful. But the fact that two responders were unnecessarily exposed still remains with me today.

—Street Story by Tom Creamer, Special Operations Coordinator, City of Worcester Fire Department, Worcester, Massachusetts

OBJECTIVES

After completing this section the student should be able to identify or explain:

■ The terms used on Material Safety Data Sheets (O)

■ The likely location of MSDSs (O)

■ Standard information available on an MSDS (O)

■ The use of the Emergency Response Guidebook (ERG) (O)

■ The type of assistance that can be provided by Chemtrec (O)

■ The methods that can be used to contact a shipper or emergency contact (O)

■ Other resources that may be available in the community (O).

INTRODUCTION

CAUTION Specific and current chemical information for the first responder is essential to protect lives and property.

Chemical information is available through a variety of sources, including those carried on emergency apparatus or information sources that can be reached via telephone, fax, or computer. The shipper and the facility are required to maintain certain documents that will assist the first responder in determining the chemical hazards they may face. Knowing what information is available and knowing how to interpret this information is a valuable tool for your protection. This chapter is divided into the most common areas of information from various sources.

EMERGENCY RESPONSE GUIDEBOOK

The *Emergency Response Guidebook* (ERG), shown in **Figure 3-2,** is a well-known book for emergency responders. It is produced by the United States Department of Transportation (DOT), Transportation Canada (TC), and Secretariat of Communications and Transportation Mexico. As the same materials are sometimes transported through these three countries, health and safety authorities decided to produce a

Figure 3-2 The DOT *Emergency Response Guidebook* is published and distributed to every emergency apparatus in the United States. It provides chemical emergency response information valuable for the first responder.

book that covered the transportation of hazardous materials (known as *dangerous goods* in Canada) for all three countries. One version is made available for every emergency response apparatus in the country. The guide is commonly referred to as the *DOT book* or the *orange book.* Updated every three years, the book contains information regarding the most commonly transported chemicals as regulated by the DOT. It is intended as a guidebook for first responders during the initial phase of a hazardous materials incident. It provides information regarding the potential hazards of responding to these materials. It is one of the only books that provides specific evacuation recommendations. Although an excellent book for first responders, it does have limitation for the more advanced responder, who requires more specific chemical information.

The book consists of these major sections:

■ Placard information

■ Listing by DOT Identification number

■ Alphabetical listing by shipping name

■ Response guides

■ Table of initial isolation and protective action distances

■ List of dangerous water reactive materials.

RESIST RUSHING IN !
APPROACH INCIDENT FROM UPWIND
STAY CLEAR OF ALL SPILLS, VAPORS, FUMES AND SMOKE

HOW TO USE THIS GUIDEBOOK DURING AN INCIDENT INVOLVING DANGEROUS GOODS

ONE   **IDENTIFY THE MATERIAL** BY FINDING ANY **ONE** OF THE FOLLOWING:

THE 4-DIGIT ID NUMBER ON A PLACARD OR ORANGE PANEL

THE 4-DIGIT ID NUMBER (after UN/NA) ON A SHIPPING DOCUMENT OR PACKAGE

THE NAME OF THE MATERIAL ON A SHIPPING DOCUMENT, PLACARD OR PACKAGE

IF AN **ID NUMBER** OR THE **NAME OF THE MATERIAL** CANNOT BE FOUND, SKIP TO THE NOTE BELOW.

TWO   **LOOK UP THE MATERIAL'S 3-DIGIT GUIDE NUMBER** IN EITHER:

THE ID NUMBER INDEX..(the yellow-bordered pages of the guidebook)

THE NAME OF MATERIAL INDEX..(the blue-bordered pages of the guidebook)

If the guide number is supplemented with the letter "P", it indicates that the material may undergo violent polymerization if subjected to heat or contamination.

If the index entry is highlighted, **LOOK FOR THE ID NUMBER AND NAME OF THE MATERIAL** IN THE TABLE OF INITIAL ISOLATION AND PROTECTIVE ACTION DISTANCES (the green-bordered pages). If necessary, **BEGIN PROTECTIVE ACTIONS IMMEDIATELY** (see the section on Protective Actions).

USE THE FOLLOWING GUIDES FOR ALL EXPLOSIVES:

> **DIVISION 1.1 (EXPLOSIVES A) - GUIDE 112**
> **DIVISION 1.2 (EXPLOSIVES A & B) - GUIDE 112**
> **DIVISION 1.3 (EXPLOSIVES B) - GUIDE 112**
> **DIVISION 1.4 (EXPLOSIVES C) - GUIDE 114**
> **DIVISION 1.5 (BLASTING AGENTS) - GUIDE 112**
> **DIVISION 1.6 - GUIDE 112**

THREE   **TURN TO THE NUMBERED GUIDE** (the orange-bordered pages) **AND READ CAREFULLY.**

NOTE   **IF A NUMBERED GUIDE CANNOT BE OBTAINED BY FOLLOWING THE ABOVE STEPS,** AND A PLACARD CAN BE SEEN, LOCATE THE PLACARD IN THE TABLE OF PLACARDS, THEN GO TO THE 3-DIGIT GUIDE SHOWN NEXT TO THE SAMPLE PLACARD.

IF A REFERENCE TO A GUIDE CANNOT BE FOUND AND THIS INCIDENT IS BELIEVED TO INVOLVE DANGEROUS GOODS, TURN TO **GUIDE 111** NOW, AND USE IT UNTIL ADDITIONAL INFORMATION BECOMES AVAILABLE. If the shipping document lists an emergency response telephone number, call that number. If the shipping document is not available, or no emergency response telephone number is listed, IMMEDIATELY CALL the appropriate **emergency response agency listed on the inside back cover of this guidebook.** Provide as much information as possible, such as the name of the carrier (trucking company or railroad) and vehicle number.

Figure 3-3   The first page of the ERG is a step listing of how to use the guide. When not sure how to proceed, the responder should turn to this page.

On the inside cover the book has an example of a shipping paper used in truck transportation, and it provides information on how the shipping paper should be written. It also provides an example of a placard with an identification number panel. The first page, which is shown in **Figure 3-3,** is an essential page for your safety; it outlines all of the actions the first responder should take. It provides a decision tree process that advances step by step through the incident. It also has an important listing of the guides for explosives. The next few pages provide information on safety precautions and contacts to call for assistance. First responders should already know where the closest local assistance will be coming

HAZARD CLASSIFICATION SYSTEM

The hazard class of dangerous goods is indicated either by its class (or division) number or name. For a placard corresponding to the primary hazard class of a material, the hazard class or division number must be displayed in the lower corner of the placard. However, no hazard class or division number may be displayed on a placard representing the subsidiary hazard of a material. For other than Class 7 or the OXYGEN placard, text indicating a hazard (for example, "CORROSIVE") is not required. Text is shown only in the U.S. The hazard class or division number must appear on the shipping document after each shipping name.

Class 1 - Explosives
Division 1.1	Explosives with a mass explosion hazard
Division 1.2	Explosives with a projection hazard
Division 1.3	Explosives with predominantly a fire hazard
Division 1.4	Explosives with no significant blast hazard
Division 1.5	Very insensitive explosives; blasting agents
Division 1.6	Extremely insensitive detonating articles

Class 2 - Gases
Division 2.1	Flammable gases
Division 2.2	Non-flammable, non-toxic* compressed gases
Division 2.3	Gases toxic* by inhalation
Division 2.4	Corrosive gases (Canada)

Class 3 - Flammable liquids (and Combustible liquids [U.S.])

Class 4 - Flammable solids; Spontaneously combustible materials; and Dangerous when wet materials
Division 4.1	Flammable solids
Division 4.2	Spontaneously combustible materials
Division 4.3	Dangerous when wet materials

Class 5 - Oxidizers and Organic peroxides
Division 5.1	Oxidizers
Division 5.2	Organic peroxides

Class 6 - Toxic* materials and Infectious substances
Division 6.1	Toxic* materials
Division 6.2	Infectious substances

Class 7 - Radioactive materials

Class 8 - Corrosive materials

Class 9 - Miscellaneous dangerous goods
Division 9.1	Miscellaneous dangerous goods (Canada)
Division 9.2	Environmentally hazardous substances (Canada)
Division 9.3	Dangerous wastes (Canada)

* The words "poison" or "poisonous" are synonymous with the word "toxic".

Figure 3-4 The DOT lists nine classes of hazardous materials in the ERG. Occasionally the only information available is the hazard class.

from and how to request state assistance if required. The DOT book provides a contact number for federal assistance; the standard is to proceed by requesting local, state, and finally federal assistance. The contact numbers and agencies are specific to Canada, United States, and Mexico. It is important for the first responder to read the DOT book prior to an incident as it provides a large amount of background material that could not be effectively read during an emergency.

The book also provides a listing of the hazard class system, which is provided in **Figure 3-4,** and offers a reference point for the hazard classes that may be listed on placards or shipping papers.

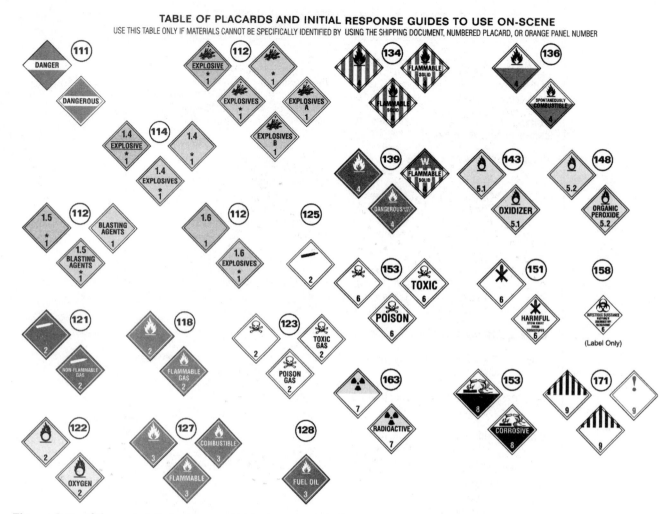

TABLE OF PLACARDS AND INITIAL RESPONSE GUIDES TO USE ON-SCENE
USE THIS TABLE ONLY IF MATERIALS CANNOT BE SPECIFICALLY IDENTIFIED BY USING THE SHIPPING DOCUMENT, NUMBERED PLACARD, OR ORANGE PANEL NUMBER

Figure 3-5 If the only information available is a placard, the ERG can provide assistance. Listed on each placard is a guide number that provides response information based on the placard information.

This listing precedes the placard section, which is valuable if the only information you have available is the placard. The placard section, which is shown in **Figure 3-5,** provides information as how to proceed at an incident where the only information is a placard. All of the possible placards are listed with their pictures, and the reader is referred to the accompanying guide page.

The yellow section, which is shown in **Figure 3-8,** is a numerical listing by the identification number or ID number. This number is assigned by the DOT to identify a material being shipped. This number takes two forms—a North America (NA) number or a United Nations (UN) number—but is always a four-digit number. Although one would think that this number would specifically identify a material, in some cases it may not. Depending on the material it may be lumped into a general category, such as Hazardous Waste, Liquid, or Not Otherwise Specified

(N.O.S.), but in most cases it will identify a specific substance. This section lists the ID number, shipping name, and guide reference number. If the material is highlighted yellow, as is shown in **Figure 3-9,** then responders must refer to the guide page listed and to the table of initial isolation and protective action distances, located at the back of the book in the green section.

Abbreviations used in the DOT ERG include:

■ N.O.S.—Not otherwise specified

■ P.I.H.—Poison inhalation hazard

■ P—Polymerization hazard

■ SCA—Surface contaminated articles (radiation)

■ LSA—Low specific activity (radiation)

■ PG I, II, or III—Packing group I, II, or III

■ O.R.M.—Other regulated material (listed as ORM A-E)

ROAD TRAILER IDENTIFICATION CHART*

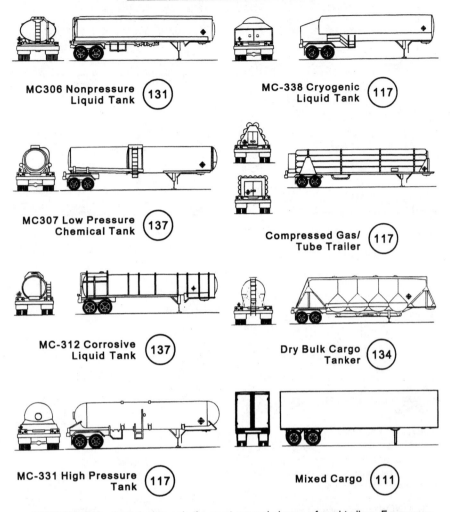

MC306 Nonpressure Liquid Tank (131)

MC-338 Cryogenic Liquid Tank (117)

MC307 Low Pressure Chemical Tank (137)

Compressed Gas/ Tube Trailer (117)

MC-312 Corrosive Liquid Tank (137)

Dry Bulk Cargo Tanker (134)

MC-331 High Pressure Tank (117)

Mixed Cargo (111)

CAUTION: This chart depicts only the most general shapes of road trailers. Emergency response personnel must be aware that there are many variations of road trailers, not illustrated above, that are used for shipping chemical products. The suggested guides are for the most hazardous products that may be transported in these trailer types.

* **The recommended guides should be considered as last resort if product cannot be identified by any other means.**

Figure 3-6a Truck silhouettes are provided with the appropriate guide page for each of the truck styles.

■ LC_{50}—Lethal concentration to 50 percent of the population

■ LD_{50}—Lethal dose to 50 percent of the population

The blue section of the book mirrors the yellow section except that it is alphabetical by shipping name, as shown in **Figure 3-10.** These shipping names are assigned by DOT and for the most part are identical to the chemical name, although they may vary in some cases. Chemicals may go by several names, including chemical name, synonym, trade name, and shipping name.

SAFETY Some chemicals can have more than 60 synonyms, which can create a confusing situation.

Table 3-1 provides one example. This section lists the ID number, shipping name, and guide reference number. If the material is highlighted blue, then responders should refer to the guide page listed and to the table of initial isolation and protective action distances, located at the back of the book in the green section.

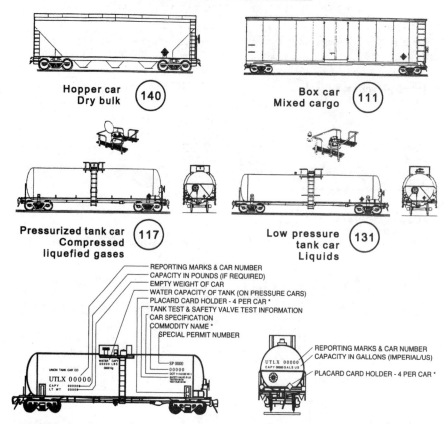

RAIL CAR IDENTIFICATION CHART*

Hopper car
Dry bulk **140**

Box car
Mixed cargo **111**

Pressurized tank car
Compressed
liquefied gases **117**

Low pressure
tank car
Liquids **131**

REPORTING MARKS & CAR NUMBER
CAPACITY IN POUNDS (IF REQUIRED)
EMPTY WEIGHT OF CAR
WATER CAPACITY OF TANK (ON PRESSURE CARS)
PLACARD CARD HOLDER - 4 PER CAR *
TANK TEST & SAFETY VALVE TEST INFORMATION
CAR SPECIFICATION
COMMODITY NAME *
SPECIAL PERMIT NUMBER

REPORTING MARKS & CAR NUMBER
CAPACITY IN GALLONS (IMPERIAL/US)

PLACARD CARD HOLDER - 4 PER CAR *

CAUTION: Emergency response personnel must be aware that rail tank cars vary widely in construction, fittings and purpose. Tank cars could transport products that may be solids, liquids or gases. The products may be under pressure. It is essential that products be identified by consulting shipping documents or train consist or contacting dispatch centers before emergency response is initiated.

The information stenciled on the sides or ends of tank cars, as illustrated above, may be used to identify the product utilizing:

a.   the commodity name shown; or

b.   the other information shown, especially reporting marks and car number which, when supplied to a dispatch center, will facilitate the identification of the product.

*   **The recommended guides should be considered as last resort if product cannot be identified by any other means.**

Figure 3-6b   Rail car silhouettes are provided with the appropriate guide page for each of the rail styles.

The middle of the book, the orange section, is the actual guide pages, as shown in **Figure 3-11.** They are listed from guide 111 to guide 172, a total of 61 guides for the more than 4,000 chemicals listed by DOT. This generalization makes this book more valuable to initial responders and less valuable to an advanced responder, such as a HAZMAT technician. A HAZMAT technician needs more specific information, which the ERG does not provide. Each guide is on two pages and is divided into three major sections: potential hazards, public safety, and emergency response. The guide may also provide some additional informa-

tion about specific chemicals. It is important for the responder to read the entire guide and the table of initial isolation and evacuation distances, if referred to that section. The potential hazards section lists the predominant hazard on the top line. If fire is the major concern, then it will be listed on top; if health is the major concern, then it will be listed first. The information regarding the health effects provides the route of exposure and any major symptoms of exposure. It will detail other potential health and environmental concerns with fire products and runoff. The fire section is not as detailed as the health section but does provide

HAZARD IDENTIFICATION CODES
DISPLAYED ON SOME INTERMODAL CONTAINERS

Hazard identification codes, referred to as "hazard identification numbers" under European and some South American regulations, may be found in the top half of an orange panel on some intermodal bulk containers. The 4-digit identification number is in the bottom half of the orange panel.

The hazard identification code in the top half of the orange panel consists of two or three figures. In general, the figures indicate the following hazards:

2 - EMISSION OF GAS DUE TO PRESSURE OR CHEMICAL REACTION

3 - FLAMMABILITY OF LIQUIDS (VAPORS) AND GASES OR SELF-HEATING LIQUID

4 - FLAMMABILITY OF SOLIDS OR SELF-HEATING SOLID

5 - OXIDIZING (FIRE-INTENSIFYING) EFFECT

6 - TOXICITY OR RISK OF INFECTION

7 - RADIOACTIVITY

8 - CORROSIVITY

9 - RISK OF SPONTANEOUS VIOLENT REACTION

- Doubling of a figure indicates an intensification of that particular hazard (i.e. 33, 66, 88).

- Where the hazard associated with a material can be adequately indicated by a single figure, the figure is followed by a zero (i.e. 30, 40, 50).

- A hazard identification code prefixed by the letter "X" indicates that the material will react dangerously with water (i.e. X88).

Figure 3-7 Intermodal containers will have an orange panel near the placard like that shown in the above figure. The 4 digit number is the DOT ID number, and the 1 or 2 digit number above that is a classification code, providing information on the basic hazards.

some assistance in identifying potential hazards. It provides information related to the flammability of the product and will identify if the vapors will stay low to the ground or will rise in air. Depending on the product it may also describe the material's ability to float on water. Products shipped at an elevated temperature will have a notation as will those materials that have the ability to polymerize. One of the problems with the ERG is that gasoline, which is highly flammable, is under the same guide that is provided for diesel fuel, which to the fire service is combustible. Although the DOT and the NFPA use 141°F as the difference between flammable and combustible, some reference sources within the fire service still uses 100°F as the difference. There is a major difference between handling a gasoline spill and a diesel fuel spill under normal conditions. By using the guide as intended, the reader has no way of differentiating between the two products. Although one could say that taking this route is safer, as it assumes the worst-case scenario, there are times when this action is inappropriate and can lead to other problems, such as an unnecessary evacuation for a diesel fuel spill.

The public safety section provides information for initial public protection options and key issues for the safety of the responders.

ID No.	Guide No.	Name of Material	ID No.	Guide No.	Name of Material
1014	122	Oxygen and Carbon dioxide mixture	1030	115	Refrigerant gas R-152a
1014	122	Oxygen and Carbon dioxide mixture, compressed	1032	118	Dimethylamine, anhydrous
			1033	115	Dimethyl ether
1015	126	Carbon dioxide and Nitrous oxide mixture	1035	115	Ethane
			1035	115	Ethane, compressed
1015	126	Nitrous oxide and Carbon dioxide mixture	1036	118	Ethylamine
			1037	115	Ethyl chloride
1016	119	Carbon monoxide	1038	115	Ethylene, refrigerated liquid (cryogenic liquid)
1016	119	Carbon monoxide, compressed			
1017	124	Chlorine	1039	115	Ethyl methyl ether
1018	126	Chlorodifluoromethane	1039	115	Methyl ethyl ether
1018	126	Refrigerant gas R-22	1040	119	Ethylene oxide
1020	126	Chloropentafluoroethane	1040	119	Ethylene oxide with Nitrogen
1020	126	Refrigerant gas R-115	1041	115	Carbon dioxide and Ethylene oxide mixture, with more than 9% but not more than 87% Ethylene oxide
1021	126	1-Chloro-1,2,2,2-tetrafluoroethane			
1021	126	Chlorotetrafluoroethane	1041	115	Carbon dioxide and Ethylene oxide mixtures, with more than 6% Ethylene oxide
1021	126	Refrigerant gas R-124			
1022	126	Chlorotrifluoromethane	1041	115	Ethylene oxide and Carbon dioxide mixture, with more than 9% but not more than 87% Ethylene oxide
1022	126	Refrigerant gas R-13			
1023	119	Coal gas			
1023	119	Coal gas, compressed	1041	115	Ethylene oxide and Carbon dioxide mixtures, with more than 6 % Ethylene oxide
1026	119	Cyanogen			
1026	119	Cyanogen, liquefied	1043	125	Fertilizer, ammoniating solution with free Ammonia
1026	119	Cyanogen gas			
1027	115	Cyclopropane	1044	126	Fire extinguishers with compressed gas
1027	115	Cyclopropane, liquefied			
1028	126	Dichlorodifluoromethane	1044	126	Fire extinguishers with liquefied gas
1028	126	Refrigerant gas R-12			
1029	126	Dichlorofluoromethane	1045	124	Fluorine
1029	126	Refrigerant gas R-21	1045	124	Fluorine, compressed
1030	115	1,1-Difluoroethane	1046	121	Helium
1030	115	Difluoroethane	1046	121	Helium, compressed

Figure 3-8 The yellow pages are a numeric listing of the chemicals that the DOT regulates. This listing is by four-digit United Nations/North America identification number and also provides the shipping name and the emergency response guide page information.

ID No.	Guide No.	Name of Material
2477	131	Methyl isothiocyanate
2478	155	Isocyanate solution, flammable, poisonous, n.o.s.
2478	155	Isocyanate solution, flammable, toxic, n.o.s.
2478	155	Isocyanate solutions, n.o.s.
2478	155	Isocyanates, flammable, poisonous, n.o.s.
2478	155	Isocyanates, flammable, toxic, n.o.s.
2478	155	Isocyanates, n.o.s.
2480	155	Methyl isocyanate
2481	155	Ethyl isocyanate
2482	155	n-Propyl isocyanate
2483	155	Isopropyl isocyanate
2484	155	tert-Butyl isocyanate
2485	155	n-Butyl isocyanate
2486	155	Isobutyl isocyanate
2487	155	Phenyl isocyanate
2488	155	Cyclohexyl isocyanate
2489	156	Diphenylmethane-4,4'-diisocyanate
2490	153	Dichloroisopropyl ether
2491	153	Ethanolamine
2491	153	Ethanolamine, solution
2491	153	Monoethanolamine
2493	132	Hexamethyleneimine
2495	144	Iodine pentafluoride
2496	156	Propionic anhydride
2497	153	Sodium phenolate, solid
2498	132	1,2,3,6-Tetrahydro-benzaldehyde
2501	152	1-Aziridinyl phosphine oxid (Tris)

Figure 3-9 If the listing in the yellow pages or the blue pages is highlighted, as is the listing of ID number 2481, ethyl isocyanate, then the responder must turn to the orange guide 155 and to the green section. The green section is the table of initial isolation and protective action distances.

NOTE It is the only text that provides recommendations as to how far to isolate the scene, and it provides some initial strategic goals for the first arriving company.

This section and the table of initial isolation and protective action distances, which is shown in **Figure 3-12,** make this book a necessary tool in the mitigation of a chemical release. It also recommends that responders contact the emergency contact listed on the shipping papers or, if this is not available, one of the contacts listed on the inside back cover. It provides some tactical objectives for handling radioactive substances and discusses additional considerations regarding these types of incidents.

The public safety section also lists personal protective equipment (PPE) recommendations and provides four basic PPE scenarios. It promotes the use of positive pressure SCBA, which is always a good recommendation when responding to chemical releases. The next level in severity recommends that chemical protective equipment is required and that turnout gear provides limited protection. In some cases the ERG states that turnout gear provides limited protection, without any recommendation for other PPE. We will discuss PPE in Chapter 4, and many of these same issues will be covered in that section. A final recommendation is that in some cases chemical protective equipment is suggested, but turnout gear is acceptable for fire situations.

This section provides information regarding initial evacuation distances for large spills and fires. If the material is listed in the table of initial isolation and protective actions distances, the reader will be referred to this section located at the back of the book in the green section. The isolation distances listed in the DOT book vary from a minimum of 30 feet to a maximum of 800 feet. The minimums are derived for materials that are solid or present little risk to the responder. The higher the toxicity or the risk, the greater the isolation distance. The average distance is 330 feet, which is what should be used for the establishment of an isolation zone for an unknown material. That distance in an urban setting can present some significant problems. As more information becomes available, this isolation distance can increase or decrease as needed. When dealing with explosives, the minimum distance should be 800 feet for the initial isolation distance. The isolation distances for downwind are expanded from the initial isolation distances and vary from 0.1 mile to 7 miles.

The emergency response section provides information regarding fires, spills, and first aid. The fire section is divided between small fires and large fires and suggests potential tactics that can be used for both. For both types of fires, the guide lists what type of extinguishing agent may be needed: water, foam, dry chemical, Halon, or CO_2. Specific materials may also list necessary special agents. Listed under the large fire section are some tactical consid-

Name of Material	Guide No.	ID No.	Name of Material	Guide No.	ID No.
Alkyl sulphonic acids, solid, with not more than 5% free Sulphuric acid	153	2585	Aluminum nitrate	140	1438
Alkylsulphuric acids	156	2571	Aluminum phosphate, solution	154	1760
Allethrin	151	2902	Aluminum phosphide	139	1397
Allyl acetate	131	2333	Aluminum phosphide pesticide	157	3048
Allyl alcohol	131	1098	Aluminum powder, coated	170	1309
Allylamine	131	2334	Aluminum powder, pyrophoric	135	1383
Allyl bromide	131	1099	Aluminum powder, uncoated	138	1396
Allyl chloride	131	1100	Aluminum processing by-products	138	3170
Allyl chlorocarbonate	155	1722	Aluminum remelting by-products	138	3170
Allyl chloroformate	155	1722	Aluminum resinate	133	2715
Allyl ethyl ether	131	2335	Aluminum silicon powder, uncoated	138	1398
Allyl formate	131	2336	Aluminum smelting by-products	138	3170
Allyl glycidyl ether	129	2219	Aluminum sulfate, solid	171	9078
Allyl iodide	132	1723	Aluminum sulfate, solution	154	1760
Allyl isothiocyanate, inhibited	155	1545	Aluminum sulphate, solid	171	9078
Allyl isothiocyanate, stabilized	155	1545	Aluminum sulphate, solution	154	1760
Allyltrichlorosilane, stabilized	155	1724	Amines, flammable, corrosive, n.o.s.	132	2733
Aluminum, molten	169	9260	Amines, liquid, corrosive, flammable, n.o.s.	132	2734
Aluminum alkyl halides	135	3052	Amines, liquid, corrosive, n.o.s.	153	2735
Aluminum alkyl hydrides	138	3076	Amines, solid, corrosive, n.o.s.	154	3259
Aluminum alkyls	135	3051	2-Amino-4-chlorophenol	151	2673
Aluminum borohydride	135	2870	2-Amino-5-diethylaminopentane	153	2946
Aluminum borohydride in devices	135	2870	2-Amino-4,6-dinitrophenol, wetted with not less than 20% water	113	3317
Aluminum bromide, anhydrous	137	1725	2-(2-Aminoethoxy)ethanol	154	1760
Aluminum bromide, solution	154	2580	2-(2-Aminoethoxy)ethanol	153	3055
Aluminum carbide	138	1394	N-Aminoethylpiperazine	153	2815
Aluminum chloride, anhydrous	137	1726	Aminophenols	152	2512
Aluminum chloride, solution	154	2581	Aminopropyldiethanolamine	154	1760
Aluminum dross	138	3170			
Aluminum ferrosilicon powder	139	1395			
Aluminum hydride	138	2463			

Figure 3-10 The blue pages are alphabetical by shipping name. These listings also provide the UN/NA number and the emergency response guide page.

erations and recommendations for specific substances. If the material is carried by tank truck or railcar, the ERG also provides some additional information on fighting those types of fires. This section provides some helpful hints that apply in many cases involving fires and transport containers:

■ Fight fire from a distance using unstaffed monitors.

■ Withdraw immediately in case of rising sound from venting safety device or if the tank begins to discolor.

■ Cool containers with flooding quantities of water until well after the fire is out.

The spill or leak section lists some general tactical objectives and also provides specific informa-

TABLE 3-1

Chemical Names

Many chemicals have several different names: chemical name, synonym, trade name, or manufacturer name. As an example the chemical known by the DOT as chlordane liquid is also called the following names:

Aspon-Chlordane	Octachlorodihydrodicyclopentadiene
Belt	1,2,4,5,6,7,8,8-Octachloro-2,3,3a,4,7,7a-Hexahydro-4,7-Methanoindene
CD 68	1,2,4,5,6,7,8,8-Octachloro-2,3,3a,4,7,7a-Hexahydro-4,7-Methano-1 H-Indene
Chloordaan	1,2,4,5,6,7,8,8-Octachloro-3a,4,7,7a-Hexahydro-4,7-Methylene indane
Chlordan	Octachloro-4,7-Methanohydroindane
g-Chlordan	Octachloro-4,7-Methanotetrahydroindane
Chlorindan	1,2,4,5,6,7,8,8-Octachloro-4,7-Methano-3a,4,7,7a-Tetrahydroindane
Chlor kil	1,2,4,5,6,7,8,8-Octachloro-3a,4,7,7a-Tetrahydro-4,7-Methanoindan
Chlorodane	1,2,4,5,6,7,8,8-Octachloro,4,7,7a-Tetrahydro-4,7-Methanoindane
Chlortox	1,2,4,5,6,7,8,8-Octachlor-3a,4,7,7a-Tetrathydo-4,7-endo-Methano-Indan (German)
Clordan	Octa-Klor
Clorodane	Oktaterr
Cortilan-Neu	Ortho-Klor
Dichlorochlordene	1,2,4,5,6,7,8,8-Ottochloro-3A,4,7,7A-Tetraidro-4,7-endo-Methno-indano (Italian)
Dowchlor	RCRA Waste Number U036
ENT 9,932	SD 5532
ENT 25,552-X	Shell SD-5532
HCS 3260	Synklor
Kypchlor	Tat Chlor 4
M 140	Topichlor 20
NCI-C00099	Topiclor
Niran	Topiclor 20
1,2,4,5,6,7,8,8-Octachloor-3a,4,7,7a-Tetrathydro4,7-endo-Methano-Indaan (Dutch)	Toxichlor
1,2,4,5,6,7,10,10-Octachloro-4,7,8,9-Tetrahydro-4,7-Methyleneindane	Velsicol 1068

Source: Excerpted from *Dangerous Properties of Industrial Materials*, 8th ed.

tion on some substances. This section that may lead responders into an action that may beyond their level of training if not used correctly. As an example, most of the guides recommend that readers stop a leak if they can do it without risk. The intention is that a responder trained to the operations level could close a remote shutoff valve. Instead, responders might incorrectly follow the recommendation and try and stop the leak no matter the method, a tactic that requires a HAZMAT technician to accomplish.

GUIDE 124 | GASES - TOXIC AND / OR CORROSIVE - OXIDIZING | NAERG96 NAERG96 | GASES - TOXIC AND / OR CORROSIVE - OXIDIZING | **GUIDE 124**

POTENTIAL HAZARDS

HEALTH
- TOXIC; may be fatal if inhaled or absorbed through skin.
- Fire will produce irritating, corrosive and/or toxic gases.
- Contact with gas or liquefied gas may cause burns, severe injury and/or frostbite.
- Runoff from fire control may cause pollution.

FIRE OR EXPLOSION
- Substance does not burn but will support combustion.
- Vapors from liquefied gas are initially heavier than air and spread along ground.
- These are strong oxidizers and will react vigorously or explosively with many materials including fuels.
- May ignite combustibles (wood, paper, oil, clothing, etc.).
- Some will react violently with air, moist air and/or water.
- Containers may explode when heated.
- Ruptured cylinders may rocket.

PUBLIC SAFETY
- CALL Emergency Response Telephone Number on Shipping Paper first. If Shipping Paper not available or no answer, refer to appropriate telephone number listed on the inside back cover.
- Isolate spill or leak area immediately for at least 100 to 200 meters (330 to 660 feet) in all directions.
- Keep unauthorized personnel away.
- Stay upwind.
- Many gases are heavier than air and will spread along ground and collect in low or confined areas (sewers, basements, tanks).
- Keep out of low areas.
- Ventilate closed spaces before entering.

PROTECTIVE CLOTHING
- Wear positive pressure self-contained breathing apparatus (SCBA).
- Wear chemical protective clothing which is specifically recommended by the manufacturer. It may provide little or no thermal protection.
- Structural firefighters' protective clothing is recommended for fire situations ONLY; it is not effective in spill situations.

EVACUATION
Spill
- See the Table of Initial Isolation and Protective Action Distances for highlighted substances. For non-highlighted substances, increase, in the downwind direction, as necessary, the isolation distance shown under "PUBLIC SAFETY".
Fire
- If tank, rail car or tank truck is involved in a fire, ISOLATE for 800 meters (1/2 mile) in all directions; also, consider initial evacuation for 800 meters (1/2 mile) in all directions.

EMERGENCY RESPONSE

FIRE
Small Fires: Water only; no dry chemical, CO₂ or Halon®.
- Contain fire and let burn. If fire must be fought, water spray or fog is recommended.
- Do not get water inside containers.
- Move containers from fire area if you can do it without risk.
- Damaged cylinders should be handled only by specialists.
Fire involving Tanks
- Fight fire from maximum distance or use unmanned hose holders or monitor nozzles.
- Cool containers with flooding quantities of water until well after fire is out.
- Do not direct water at source of leak or safety devices; icing may occur.
- Withdraw immediately in case of rising sound from venting safety devices or discoloration of tank.
- ALWAYS stay away from the ends of tanks.
- For massive fire, use unmanned hose holders or monitor nozzles; if this is impossible, withdraw from area and let fire burn.

SPILL OR LEAK
- Fully encapsulating, vapor protective clothing should be worn for spills and leaks with no fire.
- Do not touch or walk through spilled material.
- Keep combustibles (wood, paper, oil, etc.) away from spilled material.
- Stop leak if you can do it without risk.
- Use water spray to reduce vapors or divert vapor cloud drift.
- Do not direct water at spill or source of leak.
- If possible, turn leaking containers so that gas escapes rather than liquid.
- Prevent entry into waterways, sewers, basements or confined areas.
- Isolate area until gas has dispersed.
- Ventilate the area.

FIRST AID
- Move victim to fresh air.   • Call emergency medical care.
- Apply artificial respiration if victim is not breathing.
- Do not use mouth-to-mouth method if victim ingested or inhaled the substance; induce artificial respiration with the aid of a pocket mask equipped with a one-way valve or other proper respiratory medical device.
- Administer oxygen if breathing is difficult.
- Clothing frozen to the skin should be thawed before being removed.
- Remove and isolate contaminated clothing and shoes.
- In case of contact with substance, immediately flush skin or eyes with running water for at least 20 minutes.
- Keep victim warm and quiet.   • Keep victim under observation.
- Effects of contact or inhalation may be delayed.
- Ensure that medical personnel are aware of the material(s) involved, and take precautions to protect themselves.

Figure 3-11 The response guide in the orange section provides the responder with basic information regarding health, fire, and public safety issues. Each part needs to be read completely before taking any action.

Another area of concern is the suggestion to use water spray to knock down vapors. If necessary to save lives, using a water spray is an acceptable tactic.

SAFETY The indiscriminate use of water to knock down "vapors" can create many other problems.

These problems include increasing the spill size, creating runoff problems, setting off possible reactions, and posing an additional hazard to other responders working to mitigate the leak. If necessary, as in the case to knock down anhydrous ammonia vapors that may affect a neighborhood, it is an acceptable tactic. To flow water into the back of a trailer holding a leaking drum, in which the contents have not been identified is inappropriate, could cause further problems.

The first aid section provides basic medical treatment and other information and offers basic decontamination recommendations for chemical burns. The information is elementary; if confronted with a patient, the responder should contact a local hospital or Chemtrec for further assistance.

The table of initial isolation and protective action distances is the green section in the back of the book, as shown in Figure 3-12. It provides specific isolation and evacuation distances for the materials highlighted in the yellow or blue section. This section is further subdivided between small spills and large spills, and both are divided between day and night distances. The criteria used to establish these distances includes the following information: the DOT incident database (HMIS), typical package size, typical flow rate from a ruptured package, and the release rate of vapors from a spill. The DOT chose the average day as being warm and sunny with a temperature of 95°F. The

TABLE OF INITIAL ISOLATION AND PROTECTIVE ACTION DISTANCES

ID No.	NAME OF MATERIAL	SMALL SPILLS (From a small package or small leak from a large package)				LARGE SPILLS (From a large package or from many small packages)			
		First ISOLATE in all Directions		Then PROTECT persons Downwind during-		First ISOLATE in all Directions		Then PROTECT persons Downwind during-	
		Meters (Feet)		DAY Kilometers (Miles)	NIGHT Kilometers (Miles)	Meters (Feet)		DAY Kilometers (Miles)	NIGHT Kilometers (Miles)
2420	Hexafluoroacetone	60 m	(200 ft)	0.3 km (0.2 mi)	1.0 km (0.6 mi)	215 m	(700 ft)	0.8 km (0.5 mi)	3.5 km (2.2 mi)
2421	Nitrogen trioxide	60 m	(200 ft)	0.2 km (0.1 mi)	0.5 km (0.3 mi)	155 m	(500 ft)	0.5 km (0.3 mi)	1.6 km (1.0 mi)
2438	Trimethylacetyl chloride	60 m	(200 ft)	0.2 km (0.1 mi)	0.5 km (0.3 mi)	155 m	(500 ft)	0.5 km (0.3 mi)	1.9 km (1.2 mi)
2442	Trichloroacetyl chloride	60 m	(200 ft)	0.3 km (0.2 mi)	1.0 km (0.6 mi)	215 m	(700 ft)	0.8 km (0.5 mi)	3.4 km (2.1 mi)
2474	Thiophosgene	95 m	(300 ft)	0.3 km (0.2 mi)	1.1 km (0.7 mi)	215 m	(700 ft)	1.0 km (0.6 mi)	4.2 km (2.6 mi)
2477	Methyl isothiocyanate	60 m	(200 ft)	0.2 km (0.1 mi)	0.6 km (0.4 mi)	185 m	(600 ft)	0.6 km (0.4 mi)	2.4 km (1.5 mi)
2480	Methyl isocyanate	125 m	(400 ft)	0.5 km (0.3 mi)	2.3 km (1.4 mi)	305 m	(1000 ft)	1.9 km (1.2 mi)	8.2 km (5.1 mi)
2481	Ethyl isocyanate	185 m	(600 ft)	1.3 km (0.8 mi)	6.1 km (3.8 mi)	520 m	(3000 ft)	5.0 km (3.1 mi)	11.0+ km (7.0+ mi)
2482	n-Propyl isocyanate	155 m	(500 ft)	1.3 km (0.8 mi)	5.8 km (3.6 mi)	490 m	(1600 ft)	4.7 km (2.9 mi)	11.0+ km (7.0+ mi)
2483	Isopropyl isocyanate	155 m	(500 ft)	1.3 km (0.8 mi)	5.8 km (3.6 mi)	490 m	(1600 ft)	4.7 km (2.9 mi)	11.0+ km (7.0+ mi)
2484	tert-Butyl isocyanate	155 m	(500 ft)	1.1 km (0.7 mi)	5.3 km (3.3 mi)	460 m	(1500 ft)	4.3 km (2.7 mi)	11.0+ km (7.0+ mi)
2485	n-Butyl isocyanate	155 m	(500 ft)	1.1 km (0.7 mi)	5.3 km (3.3 mi)	460 m	(1500 ft)	4.3 km (2.7 mi)	11.0+ km (7.0+ mi)
2486	Isobutyl isocyanate	155 m	(500 ft)	1.1 km (0.7 mi)	5.3 km (3.3 mi)	460 m	(1500 ft)	4.3 km (2.7 mi)	11.0+ km (7.0+ mi)
2487	Phenyl isocyanate	155 m	(500 ft)	1.1 km (0.7 mi)	4.8 km (3.0 mi)	460 m	(1500 ft)	4.0 km (2.5 mi)	11.0+ km (7.0+ mi)
2488	Cyclohexyl isocyanate	155 m	(500 ft)	1.0 km (0.6 mi)	4.7 km (2.9 mi)	460 m	(1500 ft)	3.9 km (2.4 mi)	11.0+ km (7.0+ mi)
2495	Iodine pentafluoride	DANGEROUS: When spilled in water, see list at the end of this table.							
2521	Diketene, inhibited	60 m	(200 ft)	0.2 km (0.1 mi)	0.6 km (0.4 mi)	155 m	(500 ft)	0.5 km (0.3 mi)	2.3 km (1.4 mi)
2534	Methylchlorosilane	60 m	(200 ft)	0.2 km (0.1 mi)	0.8 km (0.5 mi)	185 m	(600 ft)	0.6 km (0.4 mi)	2.9 km (1.8 mi)

Figure 3-12 The table of initial isolation and protective action distances table provides recommended isolation distances for the materials that were highlighted in the yellow or blue sections. The table is divided into small spills and large spills.

agency conducted a five-year study of the weather in sixty-one cities to establish a pattern. It was during this study that it was determined that materials will travel farther at night than during the day. The DOT also chose to use an evacuation distance that in 90 percent of incidents would be too large. In this scenario, in 10 percent of the incidents the distance would be too small. The distances are a guide for the first thirty minutes of an incident and use a typical day as defined by the DOT. The ERG is a guide to get an evacuation started; as soon as possible, air monitoring needs to be established to definitively establish evacuation distances. In addition, an emergency response software package known as CAMEO uses a vapor cloud modeling program known as ALOHA. This modeling program is generally referred to as a plume dispersion model (an example of which is

shown in **Figure 3-13**) and can assist with evaluating potential downwind evacuations, using real-time weather and data.

The DOT defines a small spill as a leaking container smaller than a 55-gallon drum or a leak from a small cylinder, while a large spill is coming from a container larger than 55 gallons or a large leak from a cylinder. Leaks from several small containers may be considered a large leak. A small leak from a large container may also be considered a small leak.

The last section included in the green section, which is shown in **Figure 3-14,** is the list of dangerous water reactive materials. This section provides the evacuation distances for these materials if they contact water; the distance varies from 0.3 mile to 6 miles. One helpful item that this section provides is the chemical that a material forms

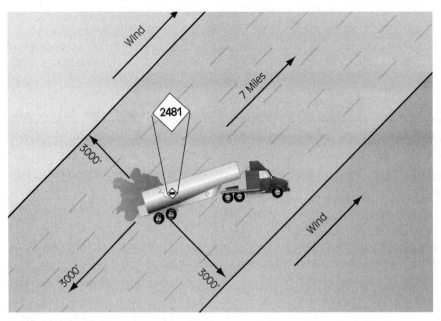

Figure 3-13 A large spill of ethyl isocyanate (ID 2481) is listed as having an isolation distance of 3,000 feet in all directions with protection for persons downwind for 7+ miles.

when contacting water, something that is unusual for this type of text.

The last pages are definitions, glossary, and explanations. The inside back cover provides a listing of additional emergency contacts for the United States, Canada, and Mexico. Further discussion regarding these agencies is found later in this chapter.

Some general reminders when using the DOT ERG:

■ It is an excellent source for evacuation and isolation distances.

■ It is an excellent source for water reactive hazards.

■ It is an excellent first responder document, one that will keep responders safe.

■ It is a great starting point for the first thirty minutes of the incident.

■ The specific chemical information is limited, and in some cases too general for a HAZMAT team, although it is well suited for the first responder level.

■ Specific identification of the product is essential for the safety of responders and citizens.

MATERIAL SAFETY DATA SHEETS

Material Safety Data Sheets (MSDSs) are a result of the hazard communication standard (HCS), which is OSHA regulation 29 CFR 1910.1200, effective 1994. As we noted in Chapter 1, this regulation requires employers who use chemicals above the household quantity to have MSDS sheets.

Employers are also required to have a hazard communication plan, label all chemical containers, and provide training to the employees on an annual basis. This training is based upon the chemical hazards the employees will face in the workplace. MSDSs are required to provide a variety of information. The amount, quality, and order of information is determined by the manufacturer. The original intent of MSDS was the protection of the employees working at the facility, not emergency responders. Although many of the sheets do have applicability and serve a useful purpose, in some cases they may not have the necessary emergency response information. The information that may be found on an MSDS is located in **Table 3-2.**

LIST OF DANGEROUS WATER-REACTIVE MATERIALS

Materials Which Create Large Amounts of Toxic (PIH) Vapor *When Spilled in Water*
* Dangerous From 0.5 to 10km (0.3 - 6.0 miles) downwind *

ID No.	Guide No.	Name of Material	Toxic Vapor (PIH) Produced	
1242	139	Methyldichlorosilane		HCl
1250	155	Methyltrichlorosilane		HCl
1295	139	Trichlorosilane		HCl
1360	139	Calcium phosphide		PH_3
1397	139	Aluminum phosphide		PH_3
1412	139	Lithium amide		NH_3
1419	139	Magnesium aluminum phosphide		PH_3
1432	139	Sodium phosphide		PH_3
1433	139	Stannic phosphides		PH_3
1471	140	Lithium hypochlorite, dry	Cl_2	HCl
1471	140	Lithium hypochlorite mixture	Cl_2	HCl
1471	140	Lithium hypochlorite mixtures, dry	Cl_2	HCl
1680	157	Potassium cyanide		HCN
1689	157	Sodium cyanide		HCN
1714	139	Zinc phosphide		PH_3
1716	156	Acetyl bromide		HBr
1717	132	Acetyl chloride		HCl
1725	137	Aluminum bromide, anhydrous		HBr
1726	137	Aluminum chloride, anhydrous		HCl
1732	157	Antimony pentafluoride		HF
1748	140	Calcium hypochlorite, dry	Cl_2	HCl
1748	140	Calcium hypochlorite mixture, dry, with more than 39% available Chlorine (8.8% available Oxygen)	Cl_2	HCl
1758	137	Chromium oxychloride		HCl
1777	137	Fluorosulfonic acid		HF
1777	137	Fluorosulphonic acid		HF

Key to PIH Formulas:

Br_2	Bromine	HF	Hydrogen fluoride	NO_2	Nitrogen dioxide
Cl_2	Chlorine	HI	Hydrogen iodide	PH_3	Phosphine
HBr	Hydrogen bromide	H_2S	Hydrogen sulfide	SO_2	Sulfur dioxide
HCl	Hydrogen chloride	H_2S	Hydrogen sulphide	SO_2	Sulphur dioxide
HCN	Hydrogen cyanide	NH_3	Ammonia		

Use this list only when material is spilled in water.

Figure 3-14 Water reactive listings found in the back of the ERG.

The quality of information varies from MSDS to MSDS and from manufacturer to manufacturer. The early MSDS sheets had a lot of good information, but as litigation has increased, the typical MSDS sheet provides a worst-case scenario and recommends an extremely conservative approach to handling the chemical. Given the choice of using the DOT ERG or a MSDS sheet, rely more on the technical information on the MSDS sheet, but responders should never rely on just one source of information. HAZMAT responders will usually follow a rule of three with regard to information sources. They will check the information and compare it with at least three sources of

CHEMICAL ABSTRACT SERVICE REGISTRY NUMBERS

With all of the synonyms and trade names, it can be very confusing trying to identify a chemical. One method is through the use of the Chemical Abstract Service (CAS) registry number, which specifically identifies a chemical. The CAS number is assigned to a chemical compound; no matter the name of the material, if it is the same compound as the original chemical, then it is assigned the same CAS number. Using Chlordane as the example, even though we have listed forty-eight different names for the compound chlordane in **Table 3-1,** they all have the same CAS number.

The majority of chemical reference sources use the CAS number, and in some cases the CAS number is used to sort MSDS. The CAS number is a series of three numbers separated by two dashes, such as xx-xxx-xxxx. Early numbers indicate that the substance was provided a number many years ago, and the numbers issued today are quite lengthy. When looking for chemical information, the CAS number is very important, as it can be used as a cross-reference source. Shipping papers and the DOT ERG do not use the CAS number, and the most common location for a CAS number is the MSDS or occasionally the label.

information, and in many cases the MSDS will be one of the sources. In June 1999 the EPA issued an alert with regard to the quality of information that MSDSs contained. In the accompanying text box we have provided most of that bulletin. The DOT ERG places many chemicals into a few general categories; although it may be conservative, the MSDS sheet is specific to the chemical. It is important through the preplanning process that responders review the chemicals at a facility with a representative from the facility. Working through the MSDS and the ERG can help determine prior to the incident which information is most useful to responders. Although many attempts have been made to modify the format of MSDS sheets and to improve the quality of information, the format has remained the same since its inception. There are a couple of recommended formats, and there exists a new format for the European chemical industry. It can be anticipated that the United States will need to conform to this format. The information provided in Table 3-2 is from this format. One of the improvements with this new MSDS format is a section specifically on emergency response procedures and considerations for responders. A sample MSDS sheet is provided in **Figure 3-15.**

SHIPPING PAPERS

In addition to the use of placards when chemicals are transported, the carrier is required to provide shipping papers for the cargo. The shipping papers in general provide the following information:

■ Shipping company
■ Destination of the packages
■ Emergency contact information
■ The number and weight of the packages
■ The name of the material or at least a hazard class of the materials
■ Special notations for hazardous materials

If the vehicle is not carrying hazardous materials, there is no requirement for any specific information, which at times can be confusing for first responders. When carrying hazardous materials the shipping papers may include a packing group number (PG) listed as I, II, or III. The lower the number, the more dangerous the chemical is; usually some special requirements accompany the shipping of the materials. Other information may include a reportable quantity (rq) for a hazardous material. Some chemicals have a threshold of reporting when spilled much like in storage situations and SARA section 304.

TABLE 3-2

Elements of a MSDS	
Information	**Note**
Chemical product and company identification	The name the chemical company uses to identify the product is used near the top of the MSDS. Information related to the manufacturer, such as the address and other contact information, is provided near the top of the first sheet. The MSDS sheet is required to list a 24-hour emergency contact number, although in most cases this number is Chemtrec.
Chemical composition	The ingredients of the chemical are listed here. If the mixture is a trade secret, there is a provision to exclude this information. If the information is needed for medical reasons the manufacturer is to provide this information to a physician. In some cases the true identity of the material may never be known, and only the hazards may be provided.
Hazards identification	In this section an emergency overview is provided, for both an employer and emergency responders. This section is lengthy and includes potential health effects, first aid, and firefighting measures. The exposure levels are typically included with the health effects, but may listed separately. New to MSDS will be a section on accidental releases, which will be beneficial to emergency responders. Handling and storage considerations and engineering controls, including proper personal protective equipment, are also provided in this section.
Physical and chemical properties	Specific chemical information, such as boiling points, vapor pressures, and flash points are included in this section.
Stability and reactivity	Some chemicals become unstable after a period of time or storage conditions, or may react with other chemicals. Any information regarding this type of information is provided here.
Toxicology information	The long term health effects and other concerns regarding acute and chronic exposures are provided in this section.
Ecological information	Information regarding any potential environmental effects are listed here.
Disposal considerations	Although in many cases this section states "follow local regulations" this is the section that outlines any regulatory requirements for disposal.
Transport information	Information regarding the DOT regulations is listed here, typically the hazard class, and UN identification number are provided in this section.
Regulatory information	Any other regulations that apply to the use, storage, or disposal of the chemical are listed here. If the chemical is covered by other regulations, this information is listed in this section.

GASES AND EQUIPMENT GROUP
MATERIAL SAFETY
DATA SHEET

SECTION 1: PRODUCT IDENTIFICATION

PRODUCT NAME: Argon, compressed
CHEMICAL NAME: Argon   **FORMULA:** Ar
SYNONYMS: Argon gas, Gaseous argon, GAR
MANUFACTURER: Air Products and Chemicals, Inc.
7201 Hamilton Boulevard
Allentown, PA 18195 - 1501
PRODUCT INFORMATION: 1 - 800 - 752 - 1597
MSDS NUMBER: 1004   **EFFECTIVE DATE:** August 1997   **REVISION:** 5

SECTION 2: COMPOSITION - INFORMATION ON INGREDIENTS

Argon is sold as pure product > 99%.
CAS NUMBER: 7440-37-1
EXPOSURE LIMITS:
  OSHA: Not established.
  ACGIH: Simple asphyxiant.
  NIOSH: Not established.

SECTION 3: HAZARDS IDENTIFICATION
EMERGENCY OVERVIEW

Argon is a nontoxic, odorless, colorless, nonflammable gas stored in cylinders at high pressure. It can cause rapid suffocation when concentrations are sufficient to reduce oxygen levels below 19.5%. It is heavier than air and may concentrate in low areas. Self Contained Breathing Apparatus (SCBA) may be required.

EMERGENCY TELEPHONE NUMBERS

800 - 523 - 9374   in Continental U.S., Canada and Puerto Rico

610 - 481 - 7711   outside U.S.

POTENTIAL HEALTH EFFECTS INFORMATION
  INHALATION: Simple asphyxiant. Argon is nontoxic, but may cause suffocation by displacing the oxygen in air. Lack of sufficient oxygen can cause serious injury or death.
  EYE CONTACT: No adverse effect.
  SKIN CONTACT: No adverse effect.
  CARCINOGENIC POTENTIAL: Argon is not listed as a carcinogen or potential carcinogen by NTP, IARC, or OSHA Subpart Z.
EXPOSURE INFORMATION
  ROUTE OF ENTRY: Inhalation

MSDS # 1004   Argon   1 of 5

TARGET ORGANS: None
EFFECT: Asphyxiation (suffocation)
MEDICAL CONDITIONS AGGRAVATED BY OVEREXPOSURE: None
SYMPTOMS: Exposure to an oxygen deficient atmosphere (<19.5%) may cause dizziness, drowsiness, nausea, vomiting, excess salivation, diminished mental alertness, loss of consciousness and death. Exposure to atmospheres containing 8-10% or less oxygen will bring about unconsciousness without warning and so quickly that the individuals cannot help themselves.

SECTION 4: FIRST AID

INHALATION: Persons suffering from lack of oxygen should be moved to fresh air. If victim is not breathing, administer artificial respiration. If breathing is difficult, administer oxygen. Obtain prompt medical attention.
EYE CONTACT: Not applicable.
SKIN CONTACT: Not applicable.

SECTION 5: FIRE AND EXPLOSION

FLASH POINT   **AUTOIGNITION TEMP**   **FLAMMABLE LIMIT**
N/A   Nonflammable   Nonflammable
EXTINGUISHING MEDIA: Argon is nonflammable and does not support combustion. Use extinguishing media appropriate for the surrounding fire.
HAZARDOUS COMBUSTION PRODUCTS: None
SPECIAL FIRE FIGHTING INSTRUCTIONS: Argon is a simple asphyxiant. If possible, remove argon cylinders from fire area or cool with water. Self contained breathing apparatus may be required for rescue workers.
UNUSUAL FIRE AND EXPLOSION HAZARDS: Upon exposure to intense heat or flame cylinder will vent rapidly and or rupture violently. Most cylinders are designed to vent contents when exposed to elevated temperatures. Pressure in a container can build up due to heat and it may rupture if pressure relief devices should fail to function.

SECTION 6: ACCIDENTAL RELEASE MEASURES

Evacuate all personnel from affected area. Increase ventilation to release area and monitor oxygen level. Use appropriate protective equipment (SCBA). If leak is from container or its valve, call the Air Products emergency telephone number. If leak is in user's system close cylinder valve and vent pressure before attempting repairs.

SECTION 7: HANDLING AND STORAGE

STORAGE: Cylinders should be stored upright in a well-ventilated, secure area, protected from the weather. Storage area temperatures should not exceed 125° F (52° C) and area should be free of combustible materials. Storage should be away from heavily traveled areas and emergency exits. Avoid areas where salt or other corrosive materials are present. Valve protection caps and valve outlet seals should remain on cylinders not connected for use. Separate full from empty cylinders. Avoid excessive inventory and storage time. Use a first-in first-out system. Keep good inventory records.
HANDLING: Do not drag, roll, or slide cylinder. Use a suitable handtruck designed for cylinder movement. Never attempt to lift a cylinder by its cap. Secure cylinders at all times while in use. Use a pressure reducing regulator or separate control valve to safely discharge gas from cylinder. Use a check valve to prevent reverse flow into cylinder. Do not overheat cylinder to increase pressure or discharge rate. If user experiences any difficulty operating cylinder valve, discontinue use and contact supplier. Never insert an object (e.g., wrench, screwdriver, pry bar, etc.) into valve cap openings. Doing so may damage valve causing a leak to occur. Use an adjustable strap-wrench to remove over-tight or rusted caps.
Argon is compatible with all common materials of construction. Pressure requirements must be considered when selecting materials and designing systems.
SPECIAL REQUIREMENTS: Always store and handle compressed gases in accordance with Compressed Gas Association, Inc. (ph.703-412-0900) pamphlet CGA P-1, *Safe Handling of Compressed Gases in Containers*. Local regulations may require specific equipment for storage or use.

MSDS # 1004   Argon   2 of 5

CAUTION: Compressed gas cylinders shall not be refilled except by qualified producers of compressed gases. Shipment of a compressed gas cylinder which has not been filled by the owner or with the owner's written consent is a violation of federal law (49 CFR 173.301).

SECTION 8: PERSONAL PROTECTION / EXPOSURE CONTROL

ENGINEERING CONTROLS: Provide good ventilation and/or local exhaust to prevent accumulation of high concentrations of gas. Oxygen levels in work area should be monitored to ensure they do not fall below 19.5%.
RESPIRATORY PROTECTION
  GENERAL USE: None required.
  EMERGENCY: Use SCBA or positive pressure air line with mask and escape pack in areas where oxygen concentration is < 19.5%. Air purifying respirators will not provide protection.
OTHER PROTECTIVE EQUIPMENT: Safety shoes and leather work gloves are recommended when handling cylinders.

SECTION 9: PHYSICAL AND CHEMICAL PROPERTIES

APPEARANCE: Colorless gas.   **ODOR:** Odorless .
MOLECULAR WEIGHT: 39.95   **BOILING POINT:** -302.2°F (-185.9°C)
SPECIFIC GRAVITY (Air =1): At 70°F (21.1°C) and 1 Atm: 1.38   **SPECIFIC VOLUME:** 9.7 ft³/lb. (0.606 m³/kg
FREEZING POINT/MELTING POINT: -308.9°F (-189.4°C)   **VAPOR PRESSURE:** Not applicable @70°F
GAS DENSITY: At 70°F (21.1°C) and 1 Atm: 0.103 lbs/ft³ (1.65 kg/m³)
SOLUBILITY IN WATER: Vol/Vol at 32 °F (0°C): 0.056

SECTION 10: STABILITY AND REACTIVITY

CHEMICAL STABILITY: Stable
CONDITIONS TO AVOID: None
INCOMPATIBILITY: None
HAZARDOUS DECOMPOSITION PRODUCTS: None
HAZARDOUS POLYMERIZATION: Will not occur.

SECTION 11: TOXICOLOGICAL INFORMATION

Argon is a simple asphyxiant.

SECTION 12: ECOLOGICAL INFORMATION

The atmosphere contains approximately 1% argon. No adverse ecological effects are expected. Argon does not contain any Class I or Class II ozone depleting chemicals. Argon is not listed as a marine pollutant by DOT (49 CFR 171).

SECTION 13: DISPOSAL

UNUSED PRODUCT / EMPTY CONTAINER: Return cylinder and unused product to supplier with the cylinder valve tightly closed and the valve caps in place. Do not attempt to dispose of residual or unused quantities.
DISPOSAL: For emergency disposal, secure the cylinder and slowly discharge gas to the atmosphere in a well ventilated area or outdoors.

MSDS # 1004   Argon   3 of 5

Figure 3-15 Material Safety
Data Sheet (*continued*).

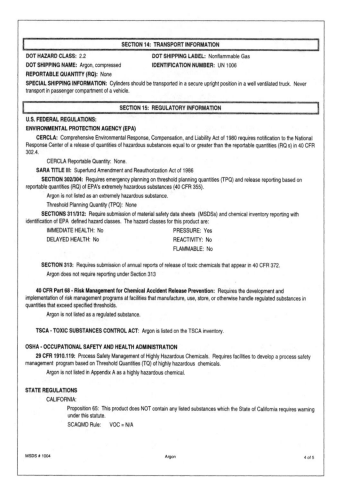

Figure 3-15 Material Safety Data Sheet *(continued)*.

The rq will be listed on the shipping papers and if the material is spilled and exceeds the quantity listed, the driver must report the spill to the **National Response Center (NRC).** The driver/operator of the vehicle is supposed to keep the shipping papers with the vehicle at all times and should be able to provide them upon request. Problems occur in accidents where there are multiple shipping papers from a variety of pickups. If the driver has kept all of the previous shipping papers from other days, it can be confusing sorting out what the actual cargo is. If the driver provides information regarding the cargo and the shipping papers agree with that information, responders should confirm the results with the shipping company. If all three are in agreement as to what is being shipped, it is probably accurate, but this is not guaranteed. A driver who picked up some illegal waste will not put it on a shipping paper or placard the vehicle; in fact, the driver probably will not admit to having it. It is obvious that the shipping company will not know about it, so until visually confirmed with

appropriate levels of PPE, responders can never be certain about a cargo's contents.

Mode of Transportation

In highway transportation the shipping papers are called a *bill of lading,* or most commonly just shipping papers. The papers are supposed to be with the driver within arms reach, most commonly in a pouch on the drivers door. On tank trucks there are occasions when a duplicate set of papers is in a tube attached near the landing gear on the driver's side. In most cases the driver will leave the papers in the truck, and they usually need to be retrieved. If drivers had many pickup stops, they will have multiple papers, and some companies put each individual item on a ticket. The hazardous materials are sometimes color coded, as in the case of United Parcel Service (UPS), which uses red tabs to identify HAZMAT packages.

For rail the shipping papers are called the **consist** or **waybill** and are in the control of the engineer or the conductor, if the train has one. There may be two

EPA ALERT

Responders should always use more than one source of information. The EPA issued a safety alert in June 1999 on the use of MSDS, and the majority of that bulletin is reproduced here.

PROBLEM: A critical consideration when choosing a response strategy is the safety of emergency responders. Adequate information about on-site chemicals can make a big difference when choosing a safe response strategy. This information must include: name, toxicity, physical and chemical characteristics, fire and reactivity hazards, emergency response procedures, spill control, and protective equipment. Generally, responders rely primarily on Material Safety Data Sheets (MSDSs) maintained at the facility. However, MSDSs may not provide sufficient information to effectively and safely respond to accidental releases. This *Alert* is designed to increase awareness of MSDS limitations, so that first responders can take proper precautions, and identify additional sources of chemical information, which could help prevent death or injury.

Accidents and How MSDSs Relate

In May 1997, a massive explosion and fire occurred at an agricultural chemical packaging facility in eastern Arkansas. Prior to the explosion, employees observed smoke in a back warehouse and evacuated. The facility called local responders and asked for help to control smoldering inside a pesticide container. The local fire department rapidly responded and reviewed the smoldering product's MSDS. The MSDS lacked information on decomposition temperatures or explosion hazards. The firefighters decided to investigate the building. While they were approaching, a violent explosion occurred. Fragments from a collapsing cinder block wall killed three firefighters and seriously injured a fourth. In April 1995, an explosion and fire at a manufacturing facility in Lodi, New Jersey, caused the death of five responders. The explosion occurred while the company was blending aluminum powder, sodium hydrosulfite, and other ingredients. Even though the material was water reactive, the MSDS for the product advised the use of a "water spray . . . to extinguish fire." The recommendation in the MSDS for "small fires" was to flood with water; however, "small fire" was not defined, the amount of water necessary was not specified, and no information dealt with how to respond to large fires (which can occur during blending processes).

The MSDS ONLY described the hazards associated with the product. In this case, responders needed information on the hazards associated with the reactivity during the blending process (which was significantly different from the product). Emergency responders should note that the chemical information provided on an MSDS usually presents the hazards associated with that particular product. Once the product is placed in a process some factors may change, resulting in the increase/decrease, or elimination of hazards. These factors may include reactions with other chemicals and changes in temperature, pressure, and physical/chemical characteristics.

MSDSs in the Workplace

In 1988, the Occupational Safety and Health Administration (OSHA) required facilities storing or using hazardous chemicals to comply with the Hazard Communication Standard. This standard requires employers to provide employees with an MSDS for every hazardous chemical present on site, and to train those employees to properly recognize the hazards of the chemicals and to handle them safely. MSDSs normally provide information on the physical/chemical characteristics and first aid procedures. This information is valuable for employees to safely work with the chemical. However, the content for MSDSs on emergency response procedures, fire, and reactive hazards may be insufficient for local responder use in an emergency situation. Vagueness, technical jargon, understandability, product vs. process concerns, and missing information on an MSDS may increase the risk to emergency responders. MSDSs are provided by manufacturers, importers, and/or distributors. MSDS chemical hazard information can vary substantially depending on the provider. Sometimes this discrepancy is due to different testing procedures. However, whoever prepared the MSDS is responsible for assuring the accuracy of the hazard information. The following chart summarizes information from various MSDSs for the chemical *azinphos methyl* and it illustrates how different sources can provide varied and conflicting information.

(continued)

EPA ALERT (continued)

Comparison of MSDS Data for Azinphos Methyl (AZM)

	MSDS–A	MSDS–B	MSDS–C	MSDS–D	CAMEO
Hazard Rating	Health—2 Flammable—0 Reactivity—0	None listed	Health—3 Flammable—2 Reactivity—2	Health—4 Flammable—0 Reactivity—0	Health—3 (extremely hazardous) Flammable—2 (ignites when moderately heated) Reactivity—2 (violent chemical change possible)
Reactivity Hazards	Stable under normal conditions. Hazardous polymerization will not occur	Depends on characteristics of dust, decomposes under influence of acids and bases	Stable material. Unstable above 100°F sustained temperature. Hazardous polymerization will not occur.	Releases toxic, corrosive, flammable, or explosive gases. Polymerization will not occur.	Will decompose
Incompatibility	High temperatures, oxidizers, alkaline substances	Acids and bases	Heat, moisture	Heat, flames, sparks, and other ignition sources	Heat, UV light
Fire Hazards	Vapors from fire are hazardous	Combustible, gives off irritating or toxic fumes (or gases) in a fire	Decomposes above 130°F with gas evolution and dense smoke. Dust explosion hazard for large dust cloud.	Containers may rupture or explode if exposed to heat.	Decomposes giving off ammonia, hydrogen, and CO.

sets of papers, one listing the contents of each car and one by car number, starting with the first car past the engine as number 1 or in some cases with the last car as number 1. In a derailment this numbering system usually goes out the window, but in some cases it may be useful. Most rail cars are identified well with car numbers and, as long as upright, can usually be identified. Rail uses a number known as a **Standard Transportation Commodity Code (STCC)**, usually referred to as a "stick" number. It is a seven-digit number, and if it starts with 49, the material is hazardous. Out of all of the modes of transportation, in rail incidents the engineer is least likely to surrender the papers and will want to accompany the papers anywhere they go. The regulations governing rail transport require the railroad to maintain procession of the papers. A system called Operation Respond is a computer program that tracks rail shipments. If you know the car number, the computer can identify what the cargo is in that particular car. The system is being slowly implemented in several cities and only involves a handful of carriers at this time.

On a ship the papers are called the **Dangerous Cargo Manifest (DCM).** These papers are in the control of the ship's captain or the master who is in second in command of the ship. No matter what the location of the ship and the crew complement, there is always someone in charge of the ship. There may be a considerable crew on aboard a ship, but the crew may only have one person who speaks English. The most difficult time to get information regarding a cargo is when the ship is loading and unloading, especially for container ships. Although they may have the papers on board for all the containers, it can be some time before crew members actually learn where each container is and what the actual cargo is. One of the major problems is learning what is in each individual container, and this information may be limited or not available at all. Luckily each container is well marked with identification information and after some time the source can be traced to get information.

NOTE One of the biggest hazards from ships comes from the fuel on board. A cruise ship can have up to a million gallons of fuel on board. When dealing with ships special training is required, as they are extremely hazardous to operate on, and responders can easily get lost and/or fall into deep areas of the ship. Another little known fact is that many ships carrying freight can also take a small number of passengers. There is a separate passenger and crew area, and the maximum number of passengers, excluding the crew, is twenty.

In air shipping the papers are called *air bills* and are in the control of the pilot, usually stored in the cockpit. The type of aircraft (passenger or cargo) will dictate the type and amount of materials that can be sent via the air. The most common aircraft involved with chemical spills are cargo aircraft, and in many cases exact identification is time consuming and difficult.

FACILITY DOCUMENTS

Each facility that has chemicals that are not consumer quantities is required to have MSDSs. In SARA Title III, facilities that are covered by these requirements are also required to submit a tier 2 form, a listing of all chemicals on site that have MSDS and a site plan. Facilities using extremely hazardous substances (EHS) are also required to have an emergency plan. Facilities may also be covered by other regulations and may be required to have other plans in place. Each facility should be able to provide upon request, relatively quickly, and without hassle a MSDS for a given material. Many facilities will have a binder of MSDSs left at the gate or with the security guard. Twenty-four-hour staff should have full access to information and have emergency contact numbers in the event of an emergency. The size of the facility will determine the amount of staff available to track this process. The SARA reports and

Figure 3-16 Most HAZMAT teams and many fire departments have chemical information on computer and can quickly and easily access this information.

information are updated annually and must be submitted by March 1 of each year. The MSDSs may not be revised annually, but the facility must ensure their accuracy. Responders should meet with the facility to review these documents as they view the facility.

COMPUTER RESOURCES

Many of the chemical information texts are also available on a computer, like that shown in **Figure 3-16,** typically a CD-ROM disk. Many first responders use **Computer Aided Management for Emergency Operations (CAMEO)** for chemical information. CAMEO is a combination software package that combines chemical response information with emergency planning capability. The information within CAMEO is easily accessible and can be used by first responders. CAMEO also has the ability to provide a vapor cloud model, known as plume dispersion, as shown in Figure 3-10. When responders input local data and leak information, a program known as Aerial Location of Hazardous Atmospheres (ALOHA) can determine the worst-case scenario for the vapor cloud travel.

CDs may also be used to store MSDS data, and many companies, especially those that are chemical distributors, have their MSDSs on the Internet. Several colleges and other large companies put their MSDSs on the Internet for public access. A simple Internet search of "Material Safety Data Sheets" provides several thousand results of possible MSDS locations. Other programs provide chemical information, but in most cases this software is specifically designed for a HAZMAT team and may be above the first responder level. One of the extreme advantages of computer software is the ability to search quickly for a chemical by its name and synonyms.

CHEMTREC

The Chemical Transportation Emergency Center (Chemtrec) is an information service provided by the American Chemistry Council. A group of chemical manufacturers established the association for several purposes, but one of the outgrowths is the Chemtrec service. This free service to emergency responders is funded by the chemical manufacturers through an annual fee. Although other services provide chemical information, Chemtrec is the largest and most experienced information service. When a company joins the Chemtrec system, it provides MSDS sheets to Chemtrec along with emergency contact information. Think of Chemtrec as a large phonebook: if you have a company name or chemical, Chemtrec can provide a contact name and number. Many shippers use Chemtrec as their emergency contact point. If Chemtrec does not have a MSDS on file, it can consult other chemical information databases. One of the big advantages is the ability to contact the manufacturer directly via conference call to get accurate information right from the product specialist. If Chemtrec does not have a specific manufacturer's name, it can connect you with other specialists who may be able to provide technical assistance. Chemtrec is well connected when providing information about chemical injuries and exposures; the service can provide medical information and a contact for further information. One thing that Chemtrec does not do is make regulatory notifications; that is the responsibility of the shipper. When calling Chemtrec, responders should have the following information available:

■ Caller's name and phone number—most responders provide their dispatch's phone number as the on-scene cell phone may be tied up.

■ Name of shipper or manufacturer—if this information is not known, then Chemtrec will contact manufacturer after manufacturer until the identity is established.

■ Shipping paper information—such as truck or railcar number, the carrier name, and the consignee (receiver) name.

■ Type of incident and local conditions.

The Canadian equivalent is called Canutec, for Canadian Transportation Emergency Center, and provides the same services as Chemtrec. In Mexico the Emergency Transportation System for the chemical industry is known as SETIQ, and it provides the same service as Chemtrec and Canutec. All three services' phone numbers and other emergency contact numbers are provided in the DOT ERG.

REFERENCE AND INFORMATION TEXTS

A lot of information texts from a variety of sources provide chemical information. At the end of this chapter is a listing of the common texts used by HAZMAT teams, like those shown in **Figure 3-17.** One recommendation is that every piece of apparatus have several reference sources. If your community has a HAZMAT team, you probably do not have additional sources of information, as contacting the

Figure 3-17 Many texts provide chemical information. No single text covers all chemicals and responders should use several references to make sure the information is available and correct.

HAZMAT team by radio or phone is quick and easy. If you are in an area where the HAZMAT team is not immediately available or has considerable travel times, then it would be prudent to have additional sources of information readily available. The DOT ERG is fairly easy to use, but many other reference sources can be complicated and may require some knowledge of chemistry terms, many of which may be above the first responder level. With some study of the preliminary information that is contained in each text, most responders can readily access the information they need. The other difference is that these texts are chemical specific and the information is for that material only. Reference texts also tend to be slanted toward the group that develops the text. Some examples are included in the Suggested Readings section provided at the end of this chapter.

A variety of texts is recommended as you can never anticipate the type of material that you may run across. If you are in an area where the HAZMAT team has an extended response time, several persons in your department should be trained to the technician level, so they can provide technical assistance to the incident commander. To function as a HAZMAT team you would need several persons trained, but one could provide chemical specific information and act as a liaison after the arrival of the HAZMAT team.

INDUSTRIAL TECHNICAL ASSISTANCE

Within each community there is usually a technical specialist in a given field. As an example, in a facility that ships and receives railcars, there is usually someone within that facility who is knowledgeable regarding railcars. In the event of an emergency in that community, the railcar specialist's expertise would be a welcome addition to the response team.

Many areas of the country, including Baltimore, Houston, and Baton Rouge, have industrial mutual aid groups designed to assist each other and the community in the event of a chemical release. In Baltimore, the South Baltimore Industrial Mutual Aid Plan (SBIMAP) provides technical specialists who assist on the scene and specialized equipment throughout a three-state area. SBIMAP has been in existence for many years, and one HAZMAT team uses the services of a chemist each time it responds to an incident.

Industrial facilities usually have employees responsible for safety and health, and those professionals are usually a good technical resource.

Within the facility there may be chemists or chemical engineers who can provide chemical information. When dealing with chemical exposures and toxicology, many facilities have industrial hygienists who work with those issues daily. As these people exist in your community, you need to search them out prior to a large incident. Determine their location and availability. When approached, many of these professionals are more than willing to assist their community.

SUMMARY

NOTE Accurate and quick information is essential to every HAZMAT incident.

The more information that can be obtained, the easier it will be to assess and respond to the incident. First responders should be starting the information process by trying to obtain as much information as possible, so they can make safe and accurate tactical decisions. If waiting for a HAZMAT team, the more information that can be obtained prior to the arrival of the team, the easier and less stressful its task will be. As essential as this information is, first responders should not take any risks attempting to get this information, such as entering a hazardous area to get shipping papers.

KEY TERMS

Air bill Term used to describe the shipping papers used in air transportation.

Computer Aided Management for Emergency Operations (CAMEO) Computer program that combines a chemical information database with emergency planning software; commonly used by HAZMAT teams to determine chemical information.

Consist Shipping papers that list the cargo of a train; the listing is by railcar of all cars.

Dangerous Cargo Manifest (DCM) Shipping papers for a ship, which list the hazardous materials on board.

National Response Center (NRC) Location a spiller is required to call to report a spill if it is in excess of the reportable quantity.

Standard Transportation Commodity Code (STCC) Number assigned to chemicals that travel by rail.

Waybill Term that may be used in conjunction with consist, but is a description of what is on a specific railcar.

REVIEW QUESTIONS

1. What are three common methods of obtaining chemical information?
2. The DOT ERG is intended to be useful for how long at an incident?
3. Given a four-digit UN identification number, which resource provides quick initial information?
4. What mode of transportation uses STCC numbers?
5. Are MSDSs required in a home for household quantities of hazardous materials?
6. Can Chemtrec be used for medical information?
7. Which section of the DOT book is used if you have a shipping name?
8. Which reference book provides information related to suggested isolation distances?
9. What advantage does an industrial contact provide at a chemical spill?
10. What cargo probably will not be listed on the shipping papers?

Suggested Readings

American Association of Railroads, *American Association of Rail Roads Hazardous Materials Action Guides,* AAR, Washington, DC, 1997. Provides basic to advanced response information; one of the few guides that actually provides cleanup and mitigation strategies; lists common materials transported by rail.

American Association of Railroads, *Emergency Handling of Hazardous Materials in Surface Transportation,* AAR, Washington, DC, 1998. Covers materials commonly transported on rail, which applies to highway incidents as well.

American Conference of Governmental Industrial Hygienists, *ACGIH Threshold Limit Values,* 1996–1997 ed. ACGIH, Cincinnati, OH. Provides the threshold limit values (TLV) for all of the chemicals that have been studied by the ACGIH; targets common industrial chemicals.

Ash, Michael and Irene (editors), *Gardener's Chemical Synonym and Trade Names,* 10th ed., Gower, 1994. Listing of common synonyms for industrial materials and household products; provided that some chemicals have more than 60 different synonyms, this text can be invaluable; no response information.

Bethrick, L. *Brethricks Handbook of Reactive Substances,* 5th ed. Butterworths' Boston, MA, 1995. One of the few sources of information on what happens when chemicals combine; not slanted toward any grouping of chemicals.

Compressed Gas Association, *Handbook of Compressed Gases* 3rd ed., Van Nostrand Reinhold, 1990. Provides extensive detail on common compressed gases; includes detailed information on each type of cylinder and common gases.

Cralley, L. J., L. V. Cralley, and R. L. Harris, (editors), *Patty's Industrial Hygiene and Toxicology series,* Volume III, Parts A and B, John Wiley & Sons, New York, NY, 1993. Detailed toxicology information resource; provides information on common industrial chemicals.

Farm Chemicals Handbook, Meister Publishing, Willoughby, OH, 1999. Definitive source for pesticide, herbicide, and insecticide information; no other text is as up-to-date or as in-depth.

Fosberg, K., and S. Z. Mansdorf, *Quick Selection Guide to Chemical Protective Clothing,* 2nd ed., Van Nostrand Reinhold, New York, NY, 1997. Provides chemical compatibility information on several hundred chemicals; listings are for the most common types of protective clothing.

Lewis, Richard J., Sr., *Hazardous Chemicals Desk Reference,* 4th ed., Van Nostrand Reinhold, New York, NY, 1996. General reference to a wide variety of hazardous materials.

Lide, David R. (editor), *Handbook of Chemistry and Physics,* 77th ed., Lewis Publishers, CRC Press, Boca Raton, FL, 1993. Detailed text on the chemical and physical properties of a hazardous material, predominantly lab chemicals; no safety information, only limited chemical and physical properties.

Merck, *MERCK Index,* 12th ed., Merck Publishing Group, Whitehouse Station, NJ, 1996. Dictionary-style reference with numerous spellings for many industrial chemicals and pharmaceuticals; provides basic chemical information, but limited response information.

National Fire Protection Association, *NFPA Fire Protection Guide on Hazardous Materials,* NFPA, Quincy, MA, 1998. Actually a combination of three texts, with sections on flammable chemicals, specific chemical hazards, and reactions.

NIOSH, *NIOSH Pocket Guide,* U.S. Department of Health and Human Services/CDC, Washington, DC, June 1977. Good source on common workplace hazards; provides information on chemical and physical properties and on protective clothing; a low-cost source of information on ionization potentials (IP), which are required when using a photoionization detector (see Chapter 6 for more information).

Noll, Gregory, Michael Hildebrand, and James Yvorra, *Hazardous Materials: Managing the Incident,* Fire Protection Publications, Oklahoma University, 1995.

Sax and Lewis, *Dangerous Properties of Industrial Materials,* 9th ed., Van Nostrand Reinhold, 1995. A useful resource, with information on chemicals used in industrial applications.

Schnepp, Rob, and Paul Gantt, *Hazardous Materials: Regulations, Response, and Site Operations,* Delmar, a division of Thomson Learning, Albany, NY, 1999.

Smeby, L. Charles (editor), *Hazardous Materials Response Handbook,* 3rd ed., National Fire Protection Association, 1997.

Stilp, Richard, and Armando Bevelacqua, *Emergency Medical Response to Hazardous Materials Incidents,* Delmar, a division of Thomson Learning, Albany, NY, 1997.

U.S. Coast Guard, *Chemical Hazards Risk Information System (CHRIS),* U.S. Coast Guard, Washington, DC, 1993. Chemicals common in water transportation, with a focus on water-based response; a great source on chemical and physical properties on land and at sea.

PROTECTION

OUTLINE

STREET STORY

Chemical protective clothing has come a long way since I was introduced to chemical hazards one evening in 1974. I had responded to a Dumpster fire early one morning. We arrived to see a thick red-brown cloud issuing from the waste container. Unfortunately we did not recognize that we had a hazardous materials situation, not a fire. Undaunted, we went about our tasks, and after half an hour, we eventually extinguished the fire. However, we all had burning eyes, nose, throat, and severe irritation to our wrists and necks. We called for the assistant chief on duty. He suggested our ailments were from the diesel fumes issuing from the fire truck. We thought otherwise but went back to the station, finished our tour, and went home.

The next morning I was visited by a man from the chemical company who wanted to make sure we were all right. He explained that the dumpster contained discarded nitric acid bottles and that some had still been full. They had started the fire. He was more concerned about our exposure, however, since we had not worn chemical protective clothing or even SCBA. (Hey, the fire was outside! Remember those days?) He went on to explain that nitric acid had a "delayed reaction to any acute exposure." This, he told me, would have affected the lungs, which might have developed fluid or edema, creating chemical pneumonia. In addition, my throat might have swollen and made it difficult to breathe. I remember saying to him, "You're probably glad I answered the door!" He agreed that it was a relief.

Over the years I have looked back at that incident, and I know it is what got me into HAZMAT. It was the driving force behind my decision to sign up for one of the first National Fire Academy Chemistry of Hazardous Materials courses. I know now that chemical protective clothing or, at the very least, SCBA should always be worn in any smoke, vapor, or cloud situation. Hindsight is the best sight, I guess.

A final note for anyone who might be concerned about me. Do not worry. Everything works out in the end. A few months later I went to an acetylene tank fire. I was the acting officer and had decided to extinguish the fire with a sodium bicarbonate extinguisher. Upon closer examination I noticed the fire was out. I turned to my engine crew to tell them to not charge the extinguisher. Sadly, one overanxious fellow discharged the unit right into my face—and me, once again without an SCBA! With my mouth wide open, I got a lung full of baking soda. So we can all relax, because I am balanced—chemically neutral—again!

—Street Story by Mike Callan, President, Callan and Company, Middlefield, Connecticut

OBJECTIVES

After completing this chapter the student should be able to identify and describe:

■ The causes of harm
■ The health hazards associated with chemical releases
■ The definitions of various chemical-related health terms
■ The various levels of PPE
■ The use of SCBA and other respiratory protection at chemical releases
■ The use of firefighting protective clothing at chemical releases
■ The signs and symptoms of heat stress.

INTRODUCTION

Personal protective equipment (PPE) takes many different shapes and versions; one example is provided in **Figure 4-2.** Even standard firefighters' turnout gear (TOG) has many variations and ensembles. When dealing with hazardous materials, these configurations are endless.

NOTE The use of PPE is essential to the health and safety of first responders.

Figure 4-2 PPE is used to protect the wearer from a variety of hazards, but no one type of PPE protects the wearer from all chemicals. This type of PPE is worn by bomb technicians to offer some protection in the event of an inadvertent detonation.

The failure to use PPE or to use it improperly is asking for an injury and may have fatal consequences.

Many responder injuries can be prevented by the proper use of PPE, most of which only takes a few seconds to put in place. The best protection for the first responder above all else is the **self-contained breathing apparatus (SCBA),** which offers a substantial amount of respiratory protection and should be considered the minimum for chemical spills. Firefighters' turnout gear does offer some limited protection against some chemicals, but it is not intended to be used for chemical spill response. TOG is not tested or approved for chemical spills. Although it may offer some protection in some chemical environments, TOG should only be used for immediate life-threatening rescue situations where there will be limited time in the hazard area. Military researchers have tested the effectiveness of TOG and SCBA in chemical warfare agent release. Agents tested included nerve agents and blister agents, such as sarin, which is a

highly toxic nerve agent. What the military found was unusual in the fact that TOG provided more protection than the chemical gear issued to soldiers. Against the chemicals that the military tested, the gear provides a high level of protection. As a matter of practice TOG should not be used in a chemical environment, with the exception of performing rescues of live victims. As the need to be in the hazard area increases, so does the need for proper PPE. This chapter outlines these types of PPE, but specific hands-on training is required prior to the use of this PPE. In general HAZMAT technicians and specialists wear chemical protective clothing, but persons trained to the operations level may be required to don chemical protective clothing to perform decon operations or other patient-related activities, after specific PPE training is provided.

HEALTH HAZARDS

Chemical exposure poses a variety of health hazards, from a short-term irritant effect to an immediate life-threatening situation to a longer-term carcinogenic risk.

Toxicology

Toxicology is the study of poisons and their effect on the body. People who study the effect of poisons on the body are known as toxicologists. Although toxicologists are typically found in the medical community, many industrial facilities employ industrial hygienists who are responsible for the protection of the workers' health and safety. Their primary focus is on the chemical hazards that exist within the facility. These persons have extensive training in toxicology and chemical exposures and are great resources to the emergency services. As the world of toxic exposures can be complicated, information is needed quickly when dealing with emergency situations. A quick consultation with an industrial hygienist may make the incident go easier.

Types of Exposures

Two types of exposures, **acute** and **chronic,** can have serious health effects.

NOTE An acute exposure is a quick one-time exposure to a chemical.

Typically, little damage or effect is noted after an acute exposure. One exception would be to some unusual acids in which a short exposure could cause minor burns. Overall, the human body does well with short duration exposures and recovers from these exposures, but all chemical exposures should be avoided. Some acute exposures may have dangerous consequences, such as being exposed to large amounts of chlorine or ammonia for a long period of time. Your body reacts adversely to these substances and has an urgent desire to evacuate any area where these chemicals may be present, usually preventing any significant health problems with these types of materials. An example of an acute exposure is a non-smoker trying a cigarette. This one-time exposure for most people is not harmful; nor would it cause any long-term health effects. By comparison, a chronic exposure is a person who has smoked three packs of cigarettes a day for twenty years. This person has received a dose of cigarette smoke several times a day for a long period of time and is most likely to have health problems associated with this chronic exposure. Abnormalities occur in both cases: It is possible that a person who smokes one cigarette may develop lung problems from that one acute exposure, just as the person who chain smokes for twenty years might never develop health problems associated with the chronic exposure.

Types of Hazards

As with the incident management systems, there are several methods of identifying possible hazards at a chemical release. The most common in use today is the acronym **TRACEM,** which stands for thermal, radiation, asphyxiation, chemical, **etiological,** and **mechanical** hazards. Each of the individual hazards have additional hazards that fit within that classification. Much like the **risk-based response** theory, the use of TRACEM puts a chemical into a risk category, and tactical decisions can be based upon that classification. The subcategories within TRACEM are as follows:

Thermal   Both heat and cold hazards fit into this category. If a flammable liquid ignites, it is classified as a thermal hazard. If liquefied oxygen contacts your skin, it would cause frostbite and a thermal (cold) burn.

Radiation   Any of the types of radiation fit into this category: alpha, beta, and gamma.

Asphyxiation   Both simple and chemical asphyxiants fit into this category.

Chemical   This category includes poisons and corrosives. The poisons may also be referred to as toxic. Some chemicals are highly toxic. There are specific levels of exposure that determine which category a chemical fits into. Also within this category are convulsants, irritants, sensitizers, and allergens. Reactions to a particular chemical vary person to person much in the way some people are allergic to bee stings while others are not. Chemicals have effects on some people while having no effect on others.

Etiological   Blood-borne pathogens and biological materials are in this category.

Mechanical   Although not chemical in nature, mechanical hazards exist within the hazard area of a chemical spill. There exists standard slip, trip, and fall hazards that one should always be concerned about. Another example of a mechanical hazard would be getting hit from blast particles, such as from a bomb or BLEVE. A drum falling on a responder is another example of a mechanical hazard.

Categories of Health Hazards

Within the TRACEM categories are several terms that responders should understand. MSDS sheets or industrial contacts may describe some chemicals as fitting into one or more of these categories. One of the most commonly used terms is **carcinogen,** which means cancer-causing potential. There are two classifications of carcinogens, known and suspected, with the majority falling into the latter category. The American Conference of Governmental Industrial Hygienists (ACGIH) provides health and safety information regarding chemical exposures, specifically cancer information.

NOTE The American Chemical Society (ACS) estimates that there are 50 million chemicals on this earth; the ACGIH states that currently only 26 have been identified as known cancer-causing agents.

There are currently only 126 suspected cancer-causing agents out of the remaining chemicals identified by the ACGIH. Among responders who deal with chemical spill response, the risk of getting cancer is always a great fear. Responders are exposed to a large number of chemicals, many of them known cancer-causing agents. As compared to the general public responders are exposed to more cancer-causing agents, but wearing PPE should prevent harm from these exposures. In prior years when SCBA was not used as extensively as it is today, cancer remains a leading cause of death among responders; many retired responders have not had the chance to enjoy retirement due to an early death.

SAFETY Wearing SCBA and PPE and avoiding other off-duty activities that involve cancer-causing materials can prevent the development of cancer in most persons.

Another commonly used term, **irritant,** is fairly self explanatory. An irritant is not corrosive but mimics the effect of a corrosive material. It can cause irritation of the eyes and possibly the respiratory tract. One notable characteristic of exposure to irritants is that the effects are easily reversed by exposure to fresh air. Mace or pepper spray are classified as irritants and may be called *incapacitating agents* by the military. **Sensitizer** is a term used to describe a

CASE STUDY

One of the common mistakes made by responders is getting tunnel vision, or focusing on only one aspect of the emergency while neglecting others. This tunnel vision can cause responders' death and injuries. My personal experience with tunnel vision that may have long-term health effects involves a known cancer-causing substance.

I arrived at work in the afternoon to begin night shift. Upon arrival at the station I learned that the daylight shift was out on a HAZMAT call in an adjacent jurisdiction. As the other night shift crew arrived, we departed to relieve the day shift on the scene. When we arrived we met with the incident commander who told us that the day shift was inside the house. They had been called to the house to investigate an unusual odor. We asked the incident commander what level of protective clothing the day shift was using. Just their uniforms, we were told. This is when tunnel vision really set in.

We decided to enter the building without speaking to any of the HAZMAT team members. When we got into the basement, we found the day shift and learned that the crew was searching for the source of an odor. There was an distinct odor that was slightly petroleum and the day shift responders said they had been unable to find the source. We asked what they had done so far and they departed for the station. As the day shift responders had said they had monitored, we did not bother getting any of the monitors. We were unable to find the source from within the house, so we decided to drill a hole in the basement and see if the source was coming from underneath the house. When the hole was finished, the odor got strong to the point of irritation. We decided to use an air monitor that the day shift had not used. The monitoring process took a while and on the last test we found the presence of benzene. The amount of benzene in the air exceeded the legal limit several times over. At that point we had been in the basement for several hours with no respiratory protection.

Ground imaging equipment was brought in the next day and located a drum buried near the house. The drum contained benzene, a known cancer-causing substance in humans. Our exposure to the benzene was extreme, and it is not known what the long-term health effects could be. The standard latency period for the development of cancer after a severe exposure to benzene is several years. The only thing I and colleagues who were there that night can do is wait and hope we don't develop cancer from this exposure. Tunnel vision got us, and I learned that I needed to evaluate the hazards no matter what the circumstances are when we arrive.

chemical effect that is in reality an allergic reaction. In most cases an employee can work with a chemical for years and suffer no effects, and then one day suffer an severe reaction to the material. Skin reddening, hives, itching, and difficulty breathing are possible symptoms when dealing with a sensitizing agent.

NOTE Some people can become sensitive to a chemical after one exposure.

Some chemicals only affect one or more organs and are described as target organ hazards, or they may affect a body system, such as the central nervous system. The effects depend on the individual, dose, concentration, and length of exposure. Some of the target organ descriptions are provided in **Table 4-1.**

Routes of Exposure

Four primary routes of exposure—respiratory, absorption, ingestion, and injection—are listed in **Figure 4-3.** The route that is the most commonly associated with causing ill health effects, both acute and chronic, is the respiratory system, as shown in **Figure 4-4.**

SAFETY In almost all cases the respiratory system requires some type of protection. SCBA will protect emergency services workers, but it must be donned and working.

The respiratory system can be affected by gases, vapors, and solid materials, such as dust and other particles. In many cases the chemicals may not have

TABLE 4-1

Target Organs and Systems	
Name	**Target organ or system**
Central nervous system (CNS)	Can cause short-term or long-term effects. Commonly short-term memory can be lost after an exposure to a CNS hazard material. Many hydrocarbons cause CNS effects, and some people purposely expose themselves to a CNS agent to receive a "high" from the exposure. In the long term, the brain cells are damaged permanently. May be referred to as neurotoxins; although they are not identical, they essentially cause the same effects.
Peripheral nervous system	Much like CNS chemicals, PNS chemicals affect the body's ability to move in a coordinated fashion. Exposure to a PNS chemical causes a disruption of the brain's ability to move messages to other body systems.
Hepatoxins	These types of materials affect the liver and, if the exposure is high enough, can cause severe liver damage.
Nephrotoxins	Affect the kidney.
Reproductive toxins	Affect the ability to reproduce or can cause birth defects. These types of toxins can stay within the body and complicate a pregnancy, even if the exposure occurred a considerable time before the pregnancy.
Mutagen	An exposure to a mutagen may not cause any effects to the people who received the exposure, but the effect could be transmitted to their offspring. Mutagens damage the genetic system and can cause mutations, which may become hereditary.
Teratogens	These materials can harm an unborn child. The effects do not happen on a cellular basis and would not be passed along from generation to generation.

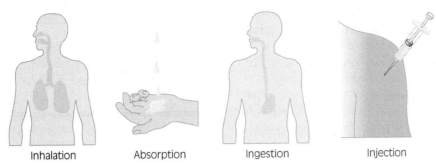

Figure 4-3 Routes of exposure.

any effect on the respiratory system itself, but may enter the body through the respiratory system and harm other organs or body systems.

There are two categories of asphyxiants, simple and chemical. Although the result is usually the same, the cause of death may differ significantly.

NOTE Simple asphyxiants exclude the oxygen in the air and push it out of the area.

It is not a chemical reaction; something else is occupying the space of the oxygen. Normal oxygen levels are 20.9 percent in air. The human body starts to develop difficulty breathing at less than 19 percent and will start to have serious problems at less than 16.5 percent oxygen. Gases such as nitrogen, halon, and carbon dioxide (CO_2) will all move oxygen out of the area and cause people in the area to have difficulty breathing; if those gases exist in high enough concentrations, death may result.

NOTE Chemical asphyxiants work in a different manner. They cause a chemical reaction within the body and will not allow it to use the readily available oxygen.

The most common chemical asphyxiant is carbon monoxide (CO). Common locations for CO exposures are listed in Chapter 6 under "Carbon Monoxide Incidents." When CO is in the air in sufficient quantities, it enters the blood stream through the lungs. It binds with the hemoglobin in the blood, forming carboxyhemoglobin. As the hemoglobin has an attraction for CO at about 225 times that of oxygen, it will not allow oxygen molecules to bind with the blood, which causes severe flulike symptoms and with, higher exposures, unconsciousness and even death.

The other common route of entry is skin absorption, as the skin is the body's largest organ. Although some chemicals can irritate the skin, this does not mean they are toxic by skin absorption. The number of chemicals that are toxic by skin

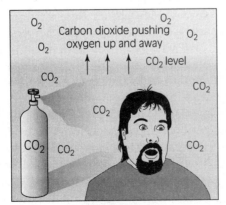

Simple Asphyxiant

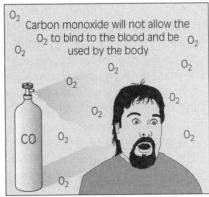

Chemical Asphyxiant

Figure 4-4 Respiratory system route of exposure.

absorption are relatively low, but precautions should be taken to minimize skin contact—even eliminate it altogether—with chemicals. Skin contact with chemicals can cause burns, rashes, or drying of the skin. The only way to provide skin protection is to wear proper protective clothing that will not allow chemicals to get onto skin. Firefighters' TOG will slow the process but will not prevent the eventual migration of the chemical to the skin. Effective decontamination is required to ensure that all of the chemical is cleaned from the skin. To best prevent chronic exposure to fire by-products, TOG should be washed on a regular basis.

Another route of entry is ingestion, which may be more common than one would think.

CAUTION After fighting a fire, responders are typically covered in soot and other debris. Most firefighters do not decontaminate themselves prior to eating or drinking refreshments that may be available at the scene. Without proper cleaning they may ingest some of these products of combustion, none of which are healthy to eat. The use of SCBA generally prevents this route of exposure, at least until the rehab phase of the incident.

One other route of exposure is injection, although it is not considered to be a major route. Many emergency services workers get exposed to hazardous materials via this route; it may be the leading route behind inhalation. The most common material that responders are exposed to are body fluids, or what is referred to as blood-borne pathogens (BBP). Exposure to BBP can occur through contact with a patients' body fluids, many times through an inadvertent needle stick. Other methods of injection are through being near a high pressure line when it breaks, such as a hydraulic rescue tool fluid line that could inject hydraulic fluid into the body. Other than standard infectious control practices there is little protection for these types of exposures except to wear full PPE properly when working in and around situations that could expose you to these fluids.

Factors that affect the rate of exposure, no matter the route, are basic items such as temperature, pulse, and respiratory rate. The higher each of these items are, the more likely the chemical will have some effect. The damage that chemicals have is based on the equation: Effect = Dose + Concentration + Time. This equation relates to acute and chronic exposures. A small dose of a small concentration for a short period of time is not likely to have any adverse effect on a normal human. A large dose at a high concentration over a long period of time will in most cases have an effect. A person who is exposed to a chemical at normal vital signs may not suffer any ill effects. But if the person jogs around the block prior to exposure, negative effects are more likely.

NOTE The increased temperature of the body under exertion will allow for faster absorption into the body, and the accompanying increased respiratory rate will cause more chemicals to enter the respiratory system.

The increased pulse rate allows for the chemicals to be spread throughout the body faster. In some confined space incidents where chemicals may have played a factor in injuries and deaths, the fact that victims' vital signs are elevated above normal allows for an increased chemical exposure and plays a role in those injuries and deaths. In most cases the first victims may be unconscious and therefore their bodies system have slowed down; in some cases they may have gone down due to a lack of oxygen. When people recognize that coworkers have gone down, their vital signs are increased and, when trying to perform a rescue, they actually get exposed to a higher level of the chemical than those they are trying to rescue. This is one of the reasons that in some cases the rescuer dies and the original victim ends up surviving.

EXPOSURE LEVELS

In industry, monitoring for exposures is commonplace and is usually intended as a preventive measure, but in the emergency services it can be an afterthought, typically after an incident has occurred. There are several different types of exposure values issued by a variety of agencies, and some of those values can be very confusing. Exposure values are established for commonly used chemicals in a variety of situations. The key to preventing exposures is to monitor for hazardous materials and to wear appropriate PPE. The one key agency involved with exposure values is the Occupational Safety and Health Administration (OSHA). The exposure values OSHA sets must be followed by all industries, including the fire service, which is considered an industry.[1] Another organization that issues exposure values is the ACGIH, which advocates worker safety and conducts a lot of studies regarding chemical exposures. The National Institute of Occupational Safety and Health (NIOSH), a research arm of OSHA, issues recommendations for exposure levels. OSHA is the only agency that provides legally binding exposure values; all of the others are recommendations. In some OSHA regulations the employer is required to use the lowest published exposure values, which in many cases are not OSHA's own values. When dealing with emergency situations, responders should always use air monitors to determine exposure levels and to follow the lowest published values. In most cases the HAZMAT team will be required to perform this function. In a response to carbon monoxide incidents, first responders may need to fulfill this role.

The exposure values are based on an average male and are for an industrial application.

NOTE The exposure values are typically based on an eight-hour day, forty-hour work week, and sixteen-hour break between exposures.

The values that are issued for the various substances are typically conservative, as in a given population it is not atypical to find someone who is sensitive to a chemical at a lower value than the rest of the group. Each value has an extra margin for safety built in, ranging from 1 to 10,000 times the actual value. The exposure values are typically listed in parts per million (PPM) or are listed as milligrams per meter cubed (mg/m$_3$); examples of these terms are provided in **Figure 4-5.**

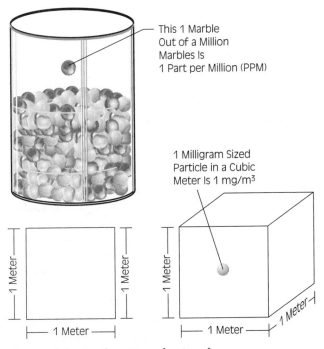

This 1 Marble Out of a Million Marbles Is 1 Part per Million (PPM)

1 Milligram Sized Particle in a Cubic Meter Is 1 mg/m³

1 Meter · 1 Meter · 1 Meter · 1 Meter · 1 Meter · 1 Meter

Figure 4-5 Explanation of units of measure.

OSHA refers to this eight-hour exposure value as the **permissible exposure limit (PEL).** The ACGIH refers to this eight-hour exposure as the **threshold limit value (TLV).** Both are average exposures over the eight-hour period. A worker can be exposed to more than the PEL or TLV, as long as at the end of the day the exposure value is less than the PEL or TLV. In some cases these exposure values are called time weighted averages (TWA), and you may see them expressed as the PEL-TWA or the TLV-TWA, which is an eight-hour-a-day average exposure. NIOSH issues **recommended exposure levels (REL),** which are recommendations for a ten-hour day forty-hour work week. These exposure values are outlined in graphic form in **Figure 4-6.**

Other values that may be listed for chemicals are the **ceiling levels,** generally referred to as PEL-C or TLV-C. These provide an amount that is the highest level an employee can be safely exposed to. When figuring an average, there are times when an employee may be exposed to chemicals at a level higher than the PEL or TLV, but there are also times when they will be exposed to less than those levels. As long as their exposure average is less than the PEL or TLV, the exposure level is acceptable. Certain chemicals will cause effects at levels that may be obtained through worker exposure and would not be considered safe, but the overall exposure would fall below the PEL or TLV. To avoid unsafe levels regulators and researchers may attach a ceiling level to an exposure value, and that is the highest level that the employee can be safely exposed to, no matter what the end average is.

Another exposure value is known as the **short term exposure limit (STEL or ST).** This is a value assigned to a fifteen-minute exposure. An employee can be exposed at this level for fifteen minutes and then is required to take an hour break from the exposure. The employee can do this four times a day without any adverse effects. NIOSH is also using an excursion value that is coupled with a time limit of five to thirty minutes typically. At this level an employee can enter an environment one time and not suffer any effects. The last value can be confusing as it is called the immediately dangerous to life or health (IDLH) value. One would think that being exposed to a chemical at the IDLH level would mean that death may be imminent. In reality this is an OSHA value that is the maximum airborne concentration that an individual could escape within thirty minutes and not suffer any adverse effects. The implied definition does not match the legal definition. At the IDLH level emergency responders need to be using SCBA as an absolute minimum, and if the chemical is toxic by skin absorption, then a fully encapsulated gas tight suit (Level A) must be used.

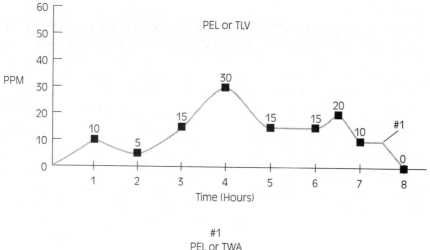

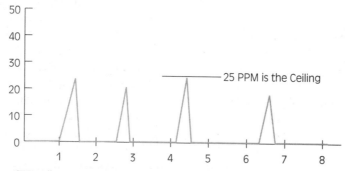

Figure 4-6 Exposure values.

Other values that a responder may see are called **lethal doses (LD$_{50}$)** or **lethal concentrations (LC$_{50}$)**. The LD is for solids and liquids, and the LC is for gases. In most cases animal studies provide these values, but there are some derived from human studies, suicides, and murders. The 50 attached to the LD or LC means the dose or concentration would be fatal to 50 percent of the exposed population. In the studies a certain number of test subjects were exposed to a low level of chemicals. After a period of time, the test subjects were studied for any adverse effects. Another group was exposed to a higher level of the chemical. When the subjects were exposed to a certain level and 50 percent of the test subjects died, this

established the LD$_{50}$ or LC$_{50}$. With threats of terrorism emergency responders must use some military data, much of which is based on LD$_{50}$ or LC$_{50}$ type studies. The military values are generally expressed as **LCt$_{50}$** or Lethal Concentration to 50 percent of the population with the t representing time, usually expressed in minutes. The military also uses **ICt$_{50}$**, which is the incapacitating concentration to 50 percent population in a certain amount of time. **Table 4-2** lists exposure values for some hazardous chemicals.

Although the exposure values can be confusing, it is important to know what each of the values mean, and emergency responders should be aware of the exposures they receive during incidents.

TABLE 4-2

Exposure Values				
Chemical	PEL	IDLH	LCt_{50} (t=3 mins.)	ICt_{50} (t=3mins.)
Sarin	0.000017 ppm	0.03 ppm	12 ppm	8 ppm
Mustard	0.0005 ppm	0.0005 ppm	231 ppm	21.5 ppm
Acetone	1000 ppm	2500 ppm	N/A	N/A
Acrolein	0.1 ppm	2 ppm	N/A	N/A
Ammonia	50 ppm (ST)	300 ppm	N/A	N/A
Chlorine	1 ppm	10 ppm	N/A	N/A
Ethion	0.02 ppm	N/A	N/A	N/A
Hydrogen sulfide	20 ppm (Ceiling)	100 ppm	N/A	N/A
Carbon monoxide	50 ppm 35 ppm NIOSH	200 (ceiling)	N/A	N/A

SAFETY Air monitoring is a key companion to these exposure levels, and it is the only way to ensure responder safety. Air monitoring for these toxic materials is a HAZMAT technician level skill.

It is generally recommended that emergency responders use the TLV or PEL as the point at which SCBA must be used. Emergency responders are faced with many chemical exposures, and at these levels safety can be ensured only if proper PPE is worn.

TYPES OF PPE

For a firefighter the most common PPE is thier turnout gear, for a law enforcement officer it is their body armor. The task determines the type of PPE, but when an law enforcement officer needs to enter an environment that presents a fire risk, then they need some type of PPE. The best type of PPE for this task is firefighters TOG, or some other fire resistive clothing. Used in this text full turnout gear is defined as helmet, hood, coat, pants, boots, gloves, PASS, and SCBA, as shown in **Figure 4-7.** For full protection, all of this equipment must be in use. All of the snaps, zippers, and closures must be used to offer optimal protection. Firefighters' TOG is designed to offer protection against heat and water. The exact amount of protection depends on the type of gear used.

The vapor barrier used in the gear plays a vital role in protection. The vapor barrier used in the military

Figure 4-7 Protective clothing, specifically the SCBA offers a high level of protection. Each chemical situation is different and the type of PPE takes careful consideration and risk assessment. Firefighters protective clothing offers protection for materials that present a fire risk.

study was a Crosstech™ vapor barrier manufactured by W. L. Gore. This vapor barrier allows the body's moisture to flow through the barrier but offers a waterproof membrane that doesn't allow water to flow back onto the skin. An unintended benefit of this vapor barrier is that it keeps chemicals out as well.

The use of firefighters' TOG is controversial as it is not certified for chemical contact. It should not be used for chemical protection, with the exception of making rescues of live victims, which should be immediately followed by emergency decontamination in many cases. Unit 5 discusses rescue issues involved with hazardous materials in more detail. Some of the latest gear is certified to protect against blood-borne pathogens, but to be sure check for NFPA certification.

Self-Contained Breathing Apparatus

> **NOTE** SCBA offers a protection factor of 10,000 from chemical exposure. In other words, a person wearing SCBA has a survivability rating 10,000 times greater than a person who is not wearing any SCBA. To offer this protection, the SCBA must be fitted properly and must be a positive pressure device.

If the SCBA fits poorly or is not a positive pressure device, the protection factor can be considerably less than 10,000. A positive pressure SCBA has an air flow in the face piece all the time and maintains a positive air pressure inside the mask to keep contaminants out. For firefighting or chemical spill response, it is imperative that the SCBA be positive pressure and that it be activated automatically without intervention of the wearer.

> **SAFETY** The odds of survival at a hazardous materials incident are dramatically improved when wearing SCBA, and it should be considered the minimum when dealing with chemical spills.

Although a variety of SCBA is available, the most common for chemical spill response is a 60-minute type, that on an average allows for a 20- to 30-minute work time for a HAZMAT team member. When determining the use of SCBA, one must consider the time it takes to enter the hazard area, working time, time to leave the area, and time to be decontaminated and undressed. In general, air supplies of 30 and 45 minutes are inadequate for spill response. Some responders to hazardous waste sites may use **supplied air respirators (SAR),** which have some advantages over SCBA. They do not have the weight of the SCBA, although the hose line necessitates some restrictions of movement. Some logistical issues are associated with the use of SARs, but for long-term incidents they are of assistance. Some response teams use **air purifying respirators (APR)** for minor spills, as shown in **Figure 4-8.** Although they offer some advantages and are common in industry, APRs are not commonly used within the emergency services. However, with terrorism and blood-borne pathogen issues becoming more commonplace, we will see an increased use of APRs. They also require **fit testing** just like the SCBA to ensure that the respirator fits and will offer the wearer protection.

Figure 4-8 This HAZMAT responder wears an air purifying respirator (APR) to filter out possible contaminants. The APR does not offer as much protection as SCBA.

Responders who choose to use an APR at an incident must follow a decision flow chart. In general, they must know the chemical, which must have good warning properties, and they have to identify the amount in the air. The amount of oxygen in the area must be verified and good levels must be maintained. A variety of cartridges can be used with APRs, and when dealing with different chemicals, a large stock must be maintained. With the exception of waste sites, minor releases, and possibly some decontamination work, SCBA is a simpler option. The only determining factor for SCBA is to find out if the air bottle is full.

Problems with SCBA include extra weight, fatigue, lack of full visibility, lack of mobility, contribution to heat stress, need to refill air supply, and other limiting factors. For this reason, many HAZMAT teams may choose the other types of respiratory protection. The type of SCBA used will determine the amount of weight added to the responder. In general the longer the work time, the higher the weight of the apparatus. This added weight adds to the overall heat stress on the user. When using SCBA the wearer has limited vision, as the face piece does not allow for a full spectrum of vision. Add a chemical protective suit and this limited field of vision gets even smaller. The use of SCBA in some applications, such as confined spaces, limits responders as to how they can move about the area. People who may be slightly claustrophobic with SCBA must cope with an additional layer of stress placed upon them when donning chemical protective clothing. The users of SCBA and chemical protective clothing should be medically cleared to function with this type of PPE, and they should receive periodic medical exams, according to the applicable regulations.

Chemical Protective Clothing

The four basic levels of chemical protective clothing are broken down into further components. The levels are assigned letters to signify their protection levels. Both OSHA and the EPA use Level A, B, C, and D, with Level A offering the highest level of chemical protection. The NFPA has established standards for protective clothing, and a description of these are provided in the accompanying text box.

Since the establishment of these levels, there have been many changes to PPE styles and types. When HAZMAT teams first started, they used reusable protective clothing that was used, cleaned, tested, and then reused. As a result of concerns over the integrity of these suits after each use, suit manufacturers decided to switch to disposable limited-use garments. Although both types of suits are available today, most teams use disposable suits. Some teams maintain at least one type of reusable garment, usually Teflon fabric suits, which are reused due to high cost.

Prior to use, all chemical protective clothing must be checked for compatibility with the chemical that is spilled.

CAUTION No one suit can be used with all chemicals.

Some NFPA-certified suits come close, but even they cannot be used with a couple of chemicals. Chemical compatibility is based on the **permeation** of a chemical through the fabric of the suit. Permeation is the movement of a chemical through the fabric on the molecular level, as shown in **Figure 4-9.**

NFPA PROTECTIVE CLOTHING STANDARDS

Although the NFPA refers to Levels A–D throughout its documents, it uses a different naming system to describe the various levels of protective clothing. OSHA/EPA Level C protective clothing equates to the NFPA 1993 Standard for Support Function Protective Garments. The standard for OSHA/EPA Level B protective clothing fits into NFPA 1992 Liquid Splash Suits for Hazardous Chemical Emergencies. As with the OSHA/EPA levels, these suits can be both encapsulating or nonencapsulating. Level A suits are outlined in NFPA 1991 Vapor Protective Suits for Hazardous Chemical Emergencies. All three standards set construction requirements and require the suits to be tested for compliance with the standard. The NFPA 1991 certified suits cost about $1,000 more than a noncertified suit, and one manufacturer reports that certified suits make up only 20 percent of its Level A sales.

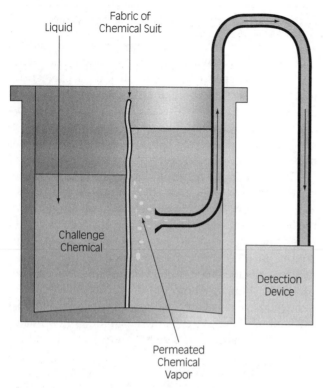

Figure 4-9 When conducting a permeation test, the fabric splits a test container, and a measurement device is used to see if the chemical goes through the fabric.

The fabric is not damaged or visibly changed in any way. Imagine wrapping garlic in plastic wrap. Initially you don't smell anything, but after a few hours the scent of garlic has permeated the refrigerator as if you hadn't wrapped it at all, because the garlic vapors have permeated the plastic wrap. Compatibility charts are provided by the manufacturer to list the chemicals the fabric of protective clothing will withstand.

Level A Protective Clothing

Level A protective clothing, as shown in **Figure 4-10,** provides the highest level of protection against chemical exposure.

NOTE For all the protection Level A gear affords responders, they remain at risk of suffering heat stress and physical and psychological stress from wearing the garment.

Level A protective clothing may also be called an **encapsulated suit.** To be rated Level A, the suit must have vapor-tight attached gloves, attached vapor-tight boots, and a gas tight zipper. Because the suit is designed to prevent gases from penetrating the garment, it is liquid tight as well. Since

materials, including air, cannot get in or out of the suit, a responder wearing it must use a SCBA the whole time. The suit does have relief valves that vent exhaled air after a certain pressure buildup.

The requirement to use Level A suits within the HAZWOPER regulation is that the atmosphere must be at or above the IDLH value, and the chemical must be toxic by skin absorption. In some cases, the use of Level A would be recommended for chemicals that would not meet that definition. Level A would be advisable in incidents in which a responder may be covered with a hazardous chemical. A good example of this situation is responders working on a valve in the bottom of a corrosive tank, especially if the valve is above their heads as they are working on it. If the valve fails, it would dump the corrosive chemical all over the responders.

SAFETY When wearing a Level A suit or any other type of PPE, pre-hydration is highly recommended.

Dressing should take place in a cool quiet area, preferably in the shade. During emergency operations all members must be monitored for heat stress. Lack of full visibility and heat stress are major concerns when using Level A protective clothing. The Level A garment typically consists of:

■ Encapsulated suit with attached gastight gloves and boots
■ Inner and outer gloves
■ Hard hat (optional)
■ Communication system (optional)
■ Cooling system (optional)
■ SCBA
■ PBI/NOMEX® coveralls (optional)
■ Overboots.

A NFPA 1991 (Encapsulated Suit Specifications) certified Level A suit also offers some flash fire resistance. To assist in compliance newer suits use a blended fabric to offer chemical resistance and flash resistance. Older style suits use a flash suit overgarment made of aluminized PBI/KEVLAR® fabric. This flash protection is not for firefighting, but it offers three to thirteen seconds of protection when exposed to a flash fire.

Level B Suits

Within the **Level B** family is a wide variety of suit types, as shown in **Figure 4-11.** Although the EPA and OSHA acknowledge two basic types, additional

PROTECTIVE CLOTHING RATINGS

Permeation is based on an eight-hour day, or 480 minutes. When a responders department's inventory has more than one suit fabric, it is best to choose a fabric with a value of >8 hours, which indicates that the fabric should show no breakthrough in eight hours. After the fabric reaches eight hours, the test is usually halted, unless specific times are indicated by the manufacturer. Most fabrics are tested against a battery of chemicals as specified by the American Society for Testing and Materials (ASTM) or the NFPA. These batteries of chemicals represent the majority of the chemical families that responders are most likely to encounter.

Two other areas of concern when dealing with PPE are degradation and penetration. The degradation of a suit involves its physical destruction, such as leaving a hole or damage. Penetration is the movement of chemicals through natural openings such as zippers, glove/suit interface points, and other areas of the suit; it does not involve damage but is an area where chemicals may enter the suit. Listed in the accompanying table are some examples of permeation data the manufacturer may provide. The suits (fabric type) are listed across the top of the table, and seven fabric types are provided in the table. Kappler Protective Fabrics provide a Level B/C suit selection called Chemical Protective Fabric (CPF) and a number 1 to 4. The Responder fabrics are usually configured in Level A suits, and the butyl and CPE fabrics are used for Level A and B suits. As you will note, no one fabric is good for all chemicals. The manufacturer tests its fabrics against hundreds of chemicals; the list provided here is only a small portion of the test results.

Chemical suit fabric (All suits™ Kappler)	CPF2	CPF3	CP4	Responder	Responder Plus	Butyl	CPE
Ammonia gas		12	>480	>480	>480		>480
Acetone	12	>480	>480	>480	>480	54	125
Carbon Disulfide		16	>480	>480	>480	8	2
Bromine				18	18		
Chlorine	>480	>480	>480	>480	>480	417	

Excerpted from http://www.kappler.com Smart Suit selection.

This table shows test results for 480 minutes (eight hours); if the suit provides protection for more than 480 minutes, it is expressed as >480 minutes. Note that the fabrics listed here provide good protection against the chemicals listed. Several suits provide adequate protection against ammonia. The CPF3 fabric offers twelve minutes of protection, while the CPF4 suit up offers >480 minutes. The remainder of the suits, except butyl, offers >480 minutes of protection. The one problem chemical is bromine, which the responder fabric can withstand for only eighteen minutes.

Keep in mind that these permeation tests put pure chemical directly against the fabric, a situation that is not normally found in the street. In most cases we do not deal with pure products. Nor are HAZMAT responders in the habit of coming in direct contact with the chemical. Another factor to consider is that most entry team times are twenty to forty minutes in length, thereby limiting the contact time with the chemicals.

---

styles are available. The two basic types are coverall style and encapsulated, but even these have subvarieties. Encapsulated Level B suits are similar to the Level A styles but are not vapor tight and do not have attached gloves. They will typically have attached booties, and the SCBA is worn on the inside. Some manufactures provide glove ring assemblies that allow for the gloves to be preattached during storage. A variety of fabrics is available for Level B suits, and compatibility is important.

NOTE The Level B encapsulated suit is the workhorse of HAZMAT teams, as it is the most common suit used.

The encapsulated Level B suit is sometimes referred to as a *Bubble B* or *B plus* suit. Although they are lighter, these suits have some of the same heat stress issues as do the Level A suits. The encapsulated Level B suit usually does not have relief valves, but one to three holes covered by a flap help

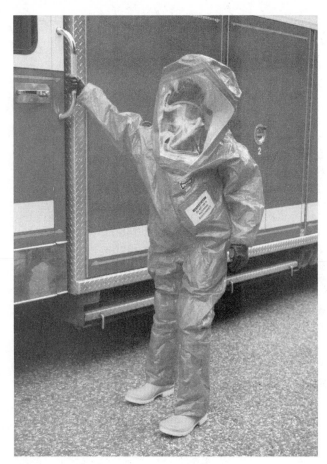

Figure 4-10 The Level A suit offers good protection from chemicals. It is a vapor-tight suit and resists permeation by most chemicals.

exhaust the exhaled air. Unlike a Level A, the Level B suit will not inflate or build pressure as this flap will not hold excess pressure. Chemicals can enter through this flap in unusual situations, whereas chemicals cannot enter through a Level A relief valve. Other styles of Level B include a two-piece garment consisting of a jacket and pants, usually bib-overall style. The coverall style Level B may have attached booties or a hood. The one item that makes the Level B different from the lower levels of PPE is the use of SCBA. Level A is a gastight suit, whereas Level B is intended for splash protection, with respiratory protection from a SCBA. A level B suit ensemble consists of:

■ Level B suit
■ Hard hat (optional)
■ Inner/outer gloves
■ SCBA
■ Communication system (optional)
■ Outer boots
■ NOMEX®/PBI coveralls (optional).

Level C Suits

Level C, as shown in **Figure 4-12,** incorporates the use of an APR within the ensemble. For obvious reasons an APR cannot be used within an encapsulated suit, but it can be used with the other styles. A Level C suit can be a coverall or a two-piece garment.

Figure 4-11 All three suits are Level B suits. From the left is a coverall style, an encapsulated suit, and two piece. Level B is intended to offer splash protection and is not gas tight.

Figure 4-12 The difference between Level C and Level B is the respirator. When using an APR, the level is C; SCBA equates to Level B. Responders cannot use an encapsulated suit and an APR, as the wearer of the APR requires good oxygen to survive.

Figure 4-13 Level D is standard work clothing and may include chemical gloves, goggles, and boots.

NOTE A Level C suit is used where splashes may occur but where respiratory hazards are minimal and are covered by the use of an APR.

When using an APR, an extensive listing of requirements must be met. A Level C garment consists of:

■ Level C suit
■ APR
■ Hard hat
■ Inner/outer gloves
■ Outer boots
■ NOMEX®/PBI coveralls.

Level D Suits

A Level D suit, shown in **Figure 4-13,** is actually work clothing. It is used when respiratory protection is not required and splashes are not a concern. Level D provides no chemical protection but offers protection against other workplace hazards. Level D protection consists of:

■ Hard hat
■ Safety glasses
■ Work clothes
■ Safety shoes/boots
■ Chemical/work gloves.

High Temperature Clothing

The two basic types of high temperature clothing are proximity and entry gear. A set of proximity gear is shown in **Figure 4-14.**

Figure 4-14 Proximity gear is used for flammable liquids firefighting, most commonly by airport firefighters and industrial firefighters.

NOTE The most common use for high temperature gear is in airport firefighting and flammable liquids firefighting.

Entry suits are sometimes used in high temperature applications within industry, such as steel making. This high temperature gear is usually identified by its characteristic aluminized outer shell. This shell is usually attached to PBI/KEVLAR® mixture, which offers a higher heat resistance than normal structural firefighting TOG. This proximity gear is named for its ability to allow the wearer in close proximity to the burning liquid. It offers protection for temperatures up to 300° to 400°F. Fire entry gear is designed to allow the wearer to enter a fully

Figure 4-15 Cryogenic gloves are used when dealing with very cold liquids. They offer protection from a range of −150 to −450°F.

involved fire area for a period of thirty to sixty seconds over the life of the suit. Its original intent was to allow for the rescue of trapped victims, but it is now well established that the victims would not be alive if that garment were needed. Useful applications for this gear involve industrial incidents where high temperatures would be found, but these are rare and most departments no longer carry this type of gear. Fire entry gear can be used in temperatures ranging up to 2000°F.

Cold Temperature Clothing

When dealing with cryogenics the responder must wear protective clothing that protects the wearer against very cold temperatures. Even propane from a cylinder, once released to the atmosphere, can reach well below 0°F. By definition, cryogenics are at least −150°F and are usually colder. Standard protective clothing will not be effective against these types of materials. Responders should take precautions against cold stress and prevent hypothermia. Although no full protective clothing ensemble is designed specifically for cold materials, various gloves, which are shown in **Figure 4-15,** gauntlets, and aprons protect against extreme cold. Layering of the clothing can offer some additional protection. Unfortunately, in addition to the cold, cryogenics usually present other hazards such as fire, corrosive, toxicity, or asphyxiation.

Limitations of PPE

Four basic limitations apply to protective clothing, including Law enforcement body armor, SWAT gear, bomb suits, EMS infectious control gear, firefighters' TOG, and fully encapsulated Level A suits. Most of these issues come to mind when dealing with chemical protective clothing, but they should be considered when using any type of protective clothing. The four major issues are heat stress, mobility, visibility, and communications problems.

Heat Stress

SAFETY Stress is a leading killer of firefighters, and heat stress plays a major factor in many of these deaths.

TOG, although getting lighter, places tremendous stress on the wearer. The type of vapor barrier also influences the amount of heat stress a person will encounter, and not all vapor barriers are created equal. It is important for emergency responders to be able to recognize heat stress and its degrees of seriousness. Heat stress can lead to heat stroke, a condition that is almost always fatal. The temperature inside a Level A suit can easily exceed 100°F, even on a cold day. When using any type of PPE, responders should take care to pre-hydrate by drinking several glasses of water and take their time when doffing their PPE. Immediately removing their PPE can shock the body, as shown in **Figure 4-16,** and cause serious health reactions. This is important to remember when removing PPE from HAZMAT responders, as they are usually zipped up inside a very warm environment and cannot control the undressing progress.

CAUTION Dehydration is a major hazard. When operating at a fire or a chemical spill, frequent hydration and rest breaks are important.

Figure 4-16 Firefighters' TOG and chemical protective clothing all place considerable stress on the responder. Steps should be taken to minimize heat stress.

The progression of heat stress is dependent on the amount of work being performed and the physical ability of the responders. To head off heat-related emergencies, responders should hydrate, as shown in **Figure 4-17,** take frequent breaks, and take extra precautions until acclimated. How acclimated responders are to heat (or cold) is indicated by the first incident on a extremely hot (or cold) day. When an incident occurs on a day in which the weather conditions would not be considered normal, responders should pay extra attention to their activity level. Early recognition is important for the health and safety of the responders. The levels and their warning signs are:

Heat Cramps   One of the initial signs of a heat-related problem is heat cramps, although some persons may never experience cramping prior to heat exhaustion. Cramping is usually in the stomach and extremities.

Heat Exhaustion   This real first step to dehydration is fairly common. The root cause is excess sweating and a loss of body fluid. Symptoms include dizziness, headache, nausea, diarrhea, vomiting, and decreased urinary output. Heat exhaustion affects not only HAZ-

Figure 4-17 Hydration is important for any emergency incident. At chemical spills it is important that the entry team hydrate prior to entering the hot sector.

MAT responders but firefighters as well. Responders should have had enough fluid that they feel a need to urinate. Untreated heat exhaustion may lead to heat stroke, which is a very serious and often fatal condition.

Heat Stroke   This is a serious emergency and can be fatal to 80 percent of the patients. In simple terms, the body's ability to regulate its temperature has failed and is no longer functioning. Symptoms include unconsciousness, hot and dry skin, seizures, confusion, and disorientation. Being dehydrated when approaching heat stroke levels only complicates the process and speeds up the patients' deterioration. It is best to prevent heat stroke, as once the body shuts down, the result is usually fatal.

Mobility

Anytime you add a layer of protective clothing, you are reducing your mobility. Each subsequent level of protective clothing places additional stress on the wearer. Part of the risk-based response profile is to make sure responders wear appropriate levels of protective clothing for the hazard. In many cases the level of protection is chosen as a knee-jerk reaction to an event. Heat stress is the major concern with PPE, but the lack of mobility adds to physiological stress. If the responder is not comfortable with the protective clothing, the potential for injury increases. With Level A clothing the suit will slightly inflate with exhaled air, creating a large balloon-like effect, until the relief valve kicks in and releases some of the internal pressure. This ballooning effect makes it difficult to move about and can create vision problems. Envision crawling on top of a railcar for the first time. The top of a railcar is not designed for a person in a Level A suit to be on top of it. A very real danger to responders on top of a railcar is the great possibility they may fall off.

Visibility

Level A and B protective suits have a small window for vision in addition to the face piece on the SCBA. This double window effect causes problems. In many cases the suit face piece will fog up, causing severe vision problems. Sometimes the application of anti-fog spray alleviates this problem, but at other times nothing helps. Another vision problem occurs when a suit is larger than it should be; the responder's head may move but the face shield on the suit does not. In these cases responders must hold the face shield of the suit and turn it with their head to see. This adds to the other stresses of a dangerous situation. When responders cannot see their hand in front of their face, the dangers increase.

Communications

The ability to communicate through a SCBA is problematic at best, and it takes a learned ear to understand many persons talking through the mask. Add a chemical suit on top of this and communication is very difficult. In most cases two persons in a chemical suit are unable to use verbal communication with each other and must rely on hand signals. Most HAZMAT teams use radio communications to maintain contact with other responders. These communications may be ear mikes in which the responder places an ear piece to hear and talk through. Other systems use a bone mike, which is strapped to the top of the skull to communicate. These systems work fairly well in most environments. One of the major problems with communications is that in some cases responders are working in a noisy environment such as a working chemical plant, which makes communications difficult.

SUMMARY

The use of protective clothing is important for the various hazards that responders may face. The absolute minimum should be fire resistive and SCBA, as this combination offers a high level of protection. If you cannot confirm the absence of hazardous materials through the use of air monitors, a basic level of protective clothing should be used. The use of chemical protective clothing is not an easy decision; many factors must be considered prior to using chemical protective clothing. Because of the ensuing heat stress that accompanies the wearing of chemical protective clothing, in some cases the protective clothing itself may be more dangerous than the chemical hazard. For this reason responders must use effective risk assessment to determine the true hazard and then dress for the hazard.

KEY TERMS

Acute   Quick, one-time exposure to a chemical.

Air purifying respirators (APR)   Respiratory protection that filters out contaminants out of the air, using filter cartridges; requires the atmosphere have sufficient oxygen along with other regulatory requirements.

Carcinogen   Material capable of causing cancer in humans.

Ceiling level   Highest exposure a person can receive without suffering any ill effects; combined with the PEL, TLV, or REL, it establishes a maximum exposure.

Chronic   Continual or repeated exposure to a hazardous material.

Encapsulated suit   Chemical suit that covers the responder, including the breathing apparatus; usually associated with gastight and liquid-tight Level A suits, but some Level B styles are fully encapsulated but not gas or liquid tight.

Etiological   Hazard that includes biological agents, viruses, and other disease-causing materials.

Fit testing   Test that ensures that the respiratory protection fits the face and offers maximum protection.

ICt_{50}   Military term for incapacitating level over set time to 50 percent of the exposed population.

Irritant   Material that is irritating to humans, but usually does not cause any long-term effects.

LCt_{50}   Military term for lethal concentration over set time to 50 percent of the population.

Level A protective clothing   Fully encapsulated chemical protective clothing; gastight and liquid-tight gear offers protection against chemical attack.

Level B   Level of protective clothing usually associated with splash protection; Level B requires the use of SCBA.

Permeation   Movement of chemicals through chemical protective clothing on a molecular level; does not cause visual damage to the clothing.

Permissible exposure limit (PEL)   OSHA value that regulates the amount of a chemical that a person can be exposed to during an eight-hour day.

Recommended exposure limit (REL)   Exposure value established by NIOSH for a ten-hour day, forty-hour workweek. Is similar to the PEL and TLV.

Risk-based response   An approach to responding to a chemical incident by categorizing a chemical into a fire, corrosive, or toxic risk. Using a risk-based approach can assist the responder in making tactical, evacuation, and PPE decisions.

Self-contained breathing apparatus (SCBA)   Protective gear that safeguards responders' respiratory systems;

Sensitizer   Chemical that after repeated exposures may cause an allergic effect for some people.

Short-term exposure limit (STEL)   Fifteen-minute exposure to a chemical; requires a one-hour break between exposures and is only allowed four times a day.

Supplied air respirators (SAR)   Respiratory protection that provides a face mask, air hose connected to a large air supply, and an escape bottle; typically used for waste sites or confined spaces.

Threshold limit value (TLV)   Exposure value that is similar to the PEL but is issued by the ACGIH. It is for an eight-hour day.

TRACEM   Acronym for the types of hazards that may exist at a chemical incident: thermal, radiation, asphyxiation, chemical, etiological, and mechanical.

REVIEW QUESTIONS

1. What are the three common routes of exposure?
2. What route of entry is the easiest to protect against?
3. To what contaminant are emergency responders commonly exposed to through injection?
4. What type of exposure is twenty years of smoking considered?
5. Which type of exposure is a one-time event?
6. What six hazards potentially exist at a chemical release?
7. Which medical effect can become hereditary?
8. Nitrogen is what type of asphyxiant?
9. Which exposure value uses a fifteen-minute time limit?

10. What is the term ceiling used for in exposure values?

11. Which level of chemical protective clothing offers a high level of protection against chemicals that are toxic through skin absorption?

12. What is a major concern as the levels of protective clothing increase?

13. Which respiratory protection offers the highest level of protection?

ENDNOTES

[1]Not all emergency services workers are covered by OSHA; some are covered by their state occupational safety agency or are not covered by such regulations. This varies state to state, and you should consult your state occupational safety office.

Suggested Readings

Lesak, David, *Hazardous Materials Strategies and Tactics,* Prentice Hall, New Jersey, 1998.

Noll, Gregory, Michael Hildebrand, and James Yvorra, *Hazardous Materials: Managing the Incident,* Fire Protection Publications, Oklahoma University, 1995.

Schnepp, Rob, and Paul Gantt, *Hazardous Materials: Regulations, Response, and Site Operations,* Delmar, a division of Thomson Learning, Albany, NY, 1999.

Smeby, L. Charles (editor), *Hazardous Materials Response Handbook,* 3rd ed., National Fire Protection Association, 1997.

CHAPTER 5

PROTECTIVE ACTIONS

OUTLINE

STREET STORY

It was the start of a normal day on the job, until later in the morning when an alarm for a chemical leak inside a beverage warehouse was sounded. The dispatch consisted of what we call a HAZMAT box: four engine companies, one truck company, a rescue squad, basic life support unit, HAZMAT company, and a commanding officer. I was working the HAZMAT company that day. While en route, a radio transmission by the first arriving company advised the communications center of a major ammonia leak inside the warehouse storage area. The first arriving company was taking protective actions. This area contained multiple storage of beverages and boxes within an enclosed and secured area that included the valves and piping for the anhydrous ammonia refrigeration system.

Upon our arrival, the warehouse had been evacuated and a strong odor of ammonia had already consumed the entire area surrounding the warehouse. Once we had performed a hazard risk assessment and ensured that the first responders had taken appropriate protective actions, we selected our level of protection. Myself and three hazardous materials technicians entered the release area to shut off the valve to the leaking pipe. After locating the release area we found the valve and made an attempt to close it. While closing the valve a sudden release of gaseous and liquid ammonia covered the personnel working at or around the valve. Visibility was taken from us almost instantaneously because of the gaseous release, and communications were lost between all four technicians. I was able to find my way out and noticed that my personnel were still in the release area. Prior to making another entry to locate my personnel, I noticed a white smoke coming from my chemical boots.

After further investigation I realized that the oil-based paint from the concrete floor was causing a chemical reaction under the soles of my boots. I reentered the release area, located my personnel, and immediately withdrew from the release area to the decontamination area. Once we were refreshed, a second entry attempt was made into the release area, where we were able to locate another sectional valve and stop the leak. The hardest part of the second attempt was removing our SCBA inside our suits to squeeze past piping and valves to get to the right one.

Before we left the scene, we finally determined that prior to our arrival a firefighter had entered the release area and closed the valve without notifying command and/or hazardous materials personnel. When hazardous materials personnel entered the release area thinking the valve was not closed, they actually reopened it, which caused the valve to freeze in the open position. In this event, a number of factors affected our response. First, arriving crews needed to address isolation and evacuation issues and the type of release, but one firefighter was endangered by not taking appropriate protective actions. No personnel were injured or exposed to the ammonia, but the incident proved to be very dangerous as a result of personnel freelancing and the lack of training present at an emergency scene.

—Street Story by Gregory Socks, Washington County Special Operations Coordinator, Hagerstown, Maryland

OBJECTIVES

After completing this chapter, the reader should be able to identify and explain:

■ The various incident management systems (A)

■ The various methods of container breaching (O)

■ The four methods of vapor cloud movement (O)

■ The methods used to determine isolation and evacuation decisions (O)

■ The use of hot, warm, and cold zones (A)

■ The use of air monitors in the determination of zones (O+)

■ Common incidents within each hazard class (O+)

■ Methods of emergency decontamination (O)

■ The four types of decontamination (O)

■ The various methods of accomplishing decontamination. (O)

INTRODUCTION

This chapter provides a myriad of topics for the responder and focuses on some general tactics that should be followed at an incident. The tactical considerations provided here are for general situations and may not apply to specific incidents. Each chemical spill is different, and for each spill there may be another way of handling that release.

NOTE The basic concept for first responders is one of isolation: Do not allow other people to become part of the incident, and protect those involved with the incident or those who may become part of the incident in a short time.

A lot of the information in this section may not apply to you specifically. As you are beginning your training, it is unlikely you will be making community evacuation decisions for the next couple of years. These are things to keep in mind in case you do have to make these decisions in the near future.

INCIDENT MANAGEMENT SYSTEMS

Several incident management systems (IMS) are designed for use in managing emergencies. The three most commonly used systems are the Incident Command System (ICS), Incident Management System (IMS), and Firescope. The Firescope system originated in California to assist in the management of forest fires. Although originally designed to be used by fire response agencies, all of the ICS or IMS systems can be used by any agency. The systems are a management system, that are used for emergencies. The basic system does not care what the emergency or disaster is, the system is a tool to manage the resources at that event. The system can be used for non emergency situations, anywhere a group is being used to manage or work an event. All of the systems are comparable in the method of managing the incident, and the focus is to enable the incident commander to manage the incident effectively. The biggest difference in the systems is some terminology. OSHA requires the use of an IMS at a hazardous materials release but does not specify which system should be used. OSHA's major concern is that all the employees know the IMS and that it is written for review by employees.

The focus of the systems is to break the incident down into manageable pieces, with officers managing a small number of people. The desired number of personnel for one person to manage, known as the **span of control,** is four to six. The systems divide the incident into five major functions, with each function having additional tasks associated with it. Each task has a leader who reports up the chain of command to the incident commander. The major tasks are:

■ **Incident commander (IC)**—The person who is in charge of the incident and is responsible for all activities at the incident. The incident commander may also have staff support at the command post who are handling:

■ Safety—Assists the incident commander with overall safety at the incident; has the authority to stop any unsafe activity without prior consultation of the incident commander. At a hazardous materials incident there are typically two safety officers, one for overall safety and one specifically assigned to the HAZMAT team to oversee specific issues.

NOTE OSHA requires the use of a safety officer and requires that the safety officer be knowledgeable about the tasks the responder will perform.

A person trained to the operations level is not adequate to be the safety officer for the

HAZMAT operation. A HAZMAT safety officer should be trained to the technician or specialist level to make sure that the mitigation efforts are carried out effectively and safely.

- ■ Public Information—The person who will meet with the press and the public to provide updates and information from the incident commander.

- ■ Liaison—There may be more than one person in this role, but the most common would be someone to interface with the police department or a police representative assigned to the command post. Other response agencies may be represented at the command post in the role of liaison.

■ **Operations**—This officer handles the tactical portion of the incident and manages the personnel who are mitigating the release. The person assigned to operations works closely with the incident commander and keeps the IC informed as to resource needs and the status of the event.

■ **Logistics**—The person responsible for obtaining and coordinating all the resources necessary to mitigate the incident. This may include just responder resources or may involve other local government or private sources of equipment. Once the equipment arrives, this person manages the equipment and provides it to operations.

■ **Finance**—The person responsible for tracking expenses involved with the incident, in addition to ensuring payment of any costs associated with the equipment that the logistics officer may obtain. In career organizations this person may track overtime and assist with the planing officer to minimize this impact. If the HAZMAT team bills the **responsible party,** the finance officer usually tracks the expenses incurred at the incident so that the responsible party can be billed for the incident.

■ **Planning**—The planning officer seeks to anticipate any needs of the incident commander. As many HAZMAT incidents have potential to last several hours to several days, this forecasting may be difficult. A good example would be when a planning officer anticipates that crews will be used for many hours and relief will be needed. Crews may also require food and refreshments, which the planning officer should anticipate. The planning officer does not request these items specifically, but runs the ideas through the incident commander who then requests these items.

The management of a hazardous materials incident can be very difficult even for a seasoned incident commander. A number of incident management strategies are available to the IC, some of which are outlined in Delmar's *Firefighter's Handbook.* HAZMAT specific systems are outlined in **Table 5-1.** A fire department IC on a normal fire incident deals predominately within one agency. There may be occasion to discuss issues with the police department, and depending on the location, a police officer may be in the command area. The IC may also deal with other municipal agencies, such as the health department or social services. Other outside agencies, such as the Red Cross, may also be involved and work under the direction of the IC. Depending on the size of the community, the media may be a factor in the management of the incident. But in the overall scheme of the incident, ICs deal mostly with their agency and their own coworkers.

When involved in a chemical release, especially one of major proportion, considerably more agencies are involved. From the start in some communities, the HAZMAT team may not be associated with the local fire department; it may be from a county mutual aid group or from an adjoining community. Many incidents involve major road closures, which bring an increased presence of police officers; depending on the size of the road, these incidents may bring command level officers to the scene. If the incident involves a state road or highway or has the potential for impacting these roads, the state highway department may arrive at the incident. If these agencies exist, county or state environmental representatives or responders may arrive at the incident to assist. In some states the Department of Natural Resources (DNR) has jurisdiction in chemical spills and may respond to an incident.

On a large spill, such as the incident depicted in **Figure 5-2,** the media will be a much larger group and may be from outside your area. At a chemical release the incident usually involves the use of a cleanup contractor who will be arriving at the incident to assist in the cleanup effort. Depending on the incident, it is not uncommon to see insurance adjusters arriving within a few hours of the spill. On larger spills or spills that may endanger a waterway, the EPA may arrive to assist with the spill. Although this listing of agencies is nowhere near complete, it represents some of the agencies that may be represented at the incident, all coming to assist the IC with mitigation. In some cases ICs may have several alarms worth of equipment from their own agencies to manage along with equipment and personnel from other agencies. Until the incident is moved from the emergency phase to the nonemergency cleanup phase the IC is usually still in

TABLE 5-1

Branch Positions

■ **Backup**—Assigned to rescue the entry team if necessary; dressed in the same suits as the entry team and fully prepared to make an entry, with the exception of being on air. The number of backup personnel needs to be a minimum of two, but can be expanded to more personnel depending on the incident. The backup team must be trained to the minimum of technician.

■ **Decontamination**—Person assigned to oversee the setup and operation of the decontamination area, sometimes referred to as the *decon area*, or the contamination reduction corridor. Other personnel will operate within the area performing the decon or PPE removal. Persons operating in this area are usually trained to the technician level, but persons trained to the operation level may also be operating in this area. If operations level personnel are used, they should have received specific training in this job function.

■ **Entry**—Two-person team at minimum who will enter the hazard area or hot zone. The type of chemical will determine the type of PPE that will be used. Prior to entry the IC must brief the personnel who will be working in the hazard area. Entry team members must be technicians or specialists.

■ **HAZMAT branch management**—Person responsible for handling the tactical objectives, as assigned by the IC. This person is in charge of the HAZMAT team and coordinates the efforts of this team; may be referred to as HAZMAT operations.

■ **HAZMAT branch safety**—Assistant to the overall safety officer, but is concerned mostly with HAZMAT specific issues. This HAZMAT safety officer is usually a technician or above and is responsible for HAZMAT specific concerns, such as PPE selection and use. The overall safety officer is concerned with all personnel at the incident and usually focuses on other hazards not related to the chemical, such as slip, trip, and fall. Slip, trip, and fall hazards are those that are caused by slippery floors or pavement, uneven terrain, or holes, and includes falls from ladders or elevated areas. These are the most common type of workplace injuries.

■ **Information/research**—Person who provides the information regarding the chemical and physical properties to the HAZMAT branch officer and the IC. After completion of these duties, this person may then assume responsibility for documenting the incident.

■ **Reconnaissance**—Through the use of binoculars or possibly by entering the hazard area, this person or team determines incident severity and identification and gathers any additional information.

■ **Resources**—Team member(s) responsible for gathering and maintaining supplies required for the incident; may be referred to as logistics.

Figure 5-2 The media plays a role in a HAZMAT incident, from relaying evacuation instructions to minimizing hysteria. A good flow of public information is essential to keep the surrounding community at ease.

charge.[1] In some cases the IC will have a more difficult time handling the "assistance" than actually managing the incident. When this occurs, public information officer (PIO) and the liaison officer will be of great assistance.

The management scenario painted in the accompanying text box does not mention victims, hospitals, or an evacuation, all of which further complicate the incident.

NOTE The use of an IMS is not only mandated by OSHA but is a good idea so that personnel can be tracked and outside agencies can be managed effectively.

IC'S LEGAL RESPONSIBILITIES

At a hazardous materials incident the IC is responsible for a number of tasks. Regardless of the local standard operating procedure, the IC has many legal obligations under the HAZWOPER regulation. The IC is responsible for the overall actions at the incident, no matter who is performing the tasks. In most cases the HAZMAT team is the technical group that performs the mitigation of the incident. Depending on the interaction of the IC and the HAZMAT team, there may not be much verbal communication. If all goes "by the book," the IC makes the decisions, but the reality is that the technical group provides the suggested options. In some cases there may be only one option. If the incident ends up as the subject of a lawsuit, the IC must answer for the HAZMAT team's actions; thus the IC must stay informed of all the incident action plans and be given the opportunity to choose an appropriate response. At a chemical release the IC is also responsible for the pre-entry briefing, which is the time to inform entry crew members of the hazards that exist and what actions are expected of them. If there are no predesignated emergency signals, these must be decided prior to entry. The IC must use a system of monitoring the progress of the incident and make changes to the system as needed. The mitigation of an incident takes the cooperation of many persons and agencies, all working toward a common goal under the guidance of the IC.

No matter what IMS is used, typically only a small component of the overall system is applied during routine fire operations. The finance branch probably will not be established during a house fire. It is likely that in a major chemical release this branch will be established and have several persons assigned to assist in this function. This applies to large and small communities with career, combination, or volunteer departments. In some fashion a liaison must be established with all of the responding agencies, and their specific roles must function within the IMS. Some specific HAZMAT branch functions may be used during an incident. These are listed in Table 5-1.

HAZARDOUS MATERIALS MANAGEMENT PROCESSES

Several different management processes can be used for hazardous materials incidents, many of which have been in use for many years and offer proven methods of organizing an incident. All of the systems have been adapted from fire service systems to fit the needs of a chemical release. The core of all these systems is basically the same, but they do differ in some areas. The core to all systems is the protection of life, property, and the environment. One IMS, the **8-step Process**® was devised by Mike Hildebrand, Greg Noll, and Jim Yvorra. Another system, designed by Dave Lesak, is the **GEDAPER**© process of hazardous materials management. Another system, developed early by Ludwig Benner, Jr., is

the **DECIDE** process, which is outlined along with the other systems in **Table 5-2.** No matter which system a department selects, it is important to use a system that everyone understands and can use. An IMS must be in place and the group of responders who are responsible for HAZMAT must fit into that IMS, using one of these processes or a combination.

Isolation and Protection

NOTE When arriving at a suspected chemical release it is important to isolate the area from other persons who may inadvertently wander into a hazardous environment.

Methods of isolation can be as simple as barrier tape, such as the scene shown in **Figure 5-3** to the use of law enforcement at traffic control points. With the increased knowledge of chemical hazards, real or otherwise, it is important to control the incident quickly. The more people who enter the suspected hazard area, the more who may later need to be rescued or, depending on the situation, require decontamination.

The protection of the persons in a hazard area can best be accomplished by evacuation of the immediate area. This does not imply that you simply tell everyone to leave, as they may need to be decontaminated or, at a minimum, medically evaluated, depending on the situation. A plan must be established for holding these persons until a deter-

CASE STUDY

The Stealth F117A fighter crash, which occurred on September 14, 1997, was an incident management challenge. The fighter crashed while performing at an air show in Baltimore County, Maryland. The fighter crashed just outside the National Guard airport, and the crash was witnessed by thousands of spectators and emergency responders. As soon as the fighter went down, crews from the airport and the county fire department responded. One of the major complications with this response was that the fighter crashed on a peninsula with only one road in and out. A large number of spectators left the air show and proceeded to the crash site. The pilot had safely ejected and was rescued by personnel from the Air Guard. The response included the Air Guard fire department, which responded with crash trucks, a first alarm assignment from the county, and a HAZMAT assignment.

The Air Guard crash trucks did an outstanding job of extinguishing the fire in and around the plane, while suppression crews were attacking multiple fires in structures. A bulk of the firefighting was done by crews who arrived from fire boats and fought the fires from the water side. The first arriving Battalion Chief Bill Purcell was faced with search and rescue, fire extinguishment in both the plane and a number of houses, EMS issues, and control of the site. Plane crashes are to be treated as crime scenes, so control of spectators and the area was essential. Because the road in was becoming clogged with spectators, the decision was made to dispatch a second alarm for additional assistance. The officers decided that extra help standing by near the scene would be better than needing help later only to find that it would be unable to make it through to the crash site.

Deputy Chief Gary Warren arrived prior to the second alarm and assumed command of the incident. The county's emergency management office was notified and was also on the scene early into the event. Many facets of the IMS were used, including operations, logistics, finance, and planning. During the investigation of the crash, residences along the peninsula were evacuated for several days. The isolation of the event initially was handled by the

county police department, and the large area required a large number of resources from the police. Early into the incident the police chief arrived to view the scene and review the activities.

Setup early on was a unified command post with the fire department assuming the role of incident commander. Other primary players in this unified command were the police department, Air Guard, Maryland Department of Environment, emergency management, and the county executive's office. The initial operations section was focused on fire control, but once the fire was knocked down, this was transitioned to the HAZMAT team to control the variety of hazards that existed. As more military personnel arrived, the military personnel changed from local air guardsmen to Air Force representatives. One of the keys to success with any incident is knowing all the players on a personal level. As this incident progressed, more and more personnel arrived who were outside of this personal level. This did have a significant impact on the event, as personnel were not familiar with each other and their agencies' policies and procedures.

As would be expected, this crash attracted considerable media attention, which taxed local resources. The Air Guard was established as the point of contact. There was substantial political attention with many local politicians stopping by to visit the site. Considerable EMS issues that developed during the incident resulted in twenty-one patients being transported. Misinformation that did not follow the chain of command may have contributed to some of these issues. It is essential that all personnel from all agencies understand and follow the IMS. The crash required the resources and presence of the Air Guard, fire department, and emergency management office for a week, all under the original IMS. The management of the outside agencies was by far the most difficult part of the event and the most taxing. When the incident went beyond a local event with the fire department and Air Guard, it grew in complexity. Although this was a small plane crash with a $46 million loss, the event was effectively handled by the IMS.

mination can be made as to their status. The HAZMAT team usually makes this determination. When dealing with people in the suspected hazard area, frequent communication with the HAZMAT team is important. The other use of protection at a chemical spill is usually associated with the adjacent community evacuation or sheltering in place. Both of these issues are further discussed later in this chapter.

Rescue

When discussing isolation and protection, rescue is a topic that naturally follows. The rescue of victims from a suspected hazard area can be extremely controversial. The decision to make a rescue is a personal one, as it may involve substantial risk to the rescuer. *Local protocol and standard operating guidelines must be considered.*

TABLE 5-2

Hazardous Materials Management Processes			
DECIDE	**8-Step Process©**	**GEDAPER©**	**HAZMAT Strategic Goals**
Detect the presence of hazardous materials	Site management and control	Gather information	Isolation
Estimate the likely harm	Identify the problem	Estimate potential course and harm	Evacuation
Choose a response objective	Hazard and risk identification	Determine strategic goals	Notification
Identify the action	Select personal protective clothing and equipment	Assess tactical options and resources	Product identification
Do the best you can	Information management and resource coordination	Plan and implement actions decided upon	Determination of appropriate personal protective equipment
Evaluate your progress	Implement response objectives	Evaluate	Decontamination
	Decontamination	Review	Spill and leak control
	Terminate the incident		Termination

The 8-Step Process© is excerpted from the text *Managing the Incident,* by Greg Noll, Mike Hildebrand, and Jim Yvorra. The GEDAPER© process is excerpted from *Hazardous Materials Strategies and Tactics,* by Dave Lesak.

SAFETY Emergency response is inherently risky. As much as we would like to eliminate all risk from our occupation, it is impossible. Our best hope is to safely manage that risk and use methods to identify that risk.

Figure 5-3 A top priority should be to isolate the area to prevent other people from becoming involved with the incident.

In reality, once responders arrive at the incident safe and sound, their chance for survival dramatically increases. One of the most dangerous parts of firefighting is responding to the incident. Each year 20 to 30 percent of firefighter deaths are the result of responding to and from an incident.

With specific regard to TOG, scientific data published in August 1999 provides ICs with some information as to the ability to make rescues. The Soldiers Biological and Chemical Command (SBCOM), formerly known as Chemical and Biological Defense Command (CBDCOM) located at the Aberdeen Proving Grounds in Maryland, performed several studies involving firefighters' protective clothing. Researchers found that firefighters protective clothing can offer protection in situations where there are significant hazards to someone without protective clothing. They tested both nerve agents and blister agents, both of which present toxicity hazards; the nerve agent is extremely toxic. In situations where a nerve agent has been released and there are both live victims and dead victims, firefighters in full PPE and SCBA can enter this environment and make rescues with no ill effects. In a situation in which all the victims are dead, firefighters can enter this severely toxic environment for three minutes with no ill effects. It goes without

THE IMPORTANCE OF ISOLATING THE SCENE

The first arriving responders can make a hazardous material release easier to manage if they begin the isolation process. In many areas a bus accident causes responders to act quickly to isolate the area and deny entry. In many cities the people on a bus that has been involved in an accident will receive a check in their names in exchange for a release from further damage claims against the bus company. Failure to control access points to the bus will result in a dramatic increase in the number of occupants, many seeking the payment from the bus company.

What holds true with the bus accident is now true with chemical releases. In many cases checks are being written or people plan to sue for possible damages, making isolation difficult. The mention of chemical exposure may result in people attempting to enter a potential hazard area. One of the most difficult areas to attempt to control is a mall or shopping center in which hundreds of people may be involved. A simple discharge of pepper spray will result in a number of real victims and usually generate a few more suspect victims. Everyone must be treated and possibly transported, which can create a strain on the EMS system. Early isolation and evacuation can help reduce the numbers of victims.

---

saying that in these cases emergency decon is a must and the gear will probably be destroyed.

When involved in making rescues at hazardous materials incidents, responders must use the same type of clues that drive making rescues in fire situations. The type of occupancy, time of day, visible clues, audible clues, and eyewitness reports are only some of the clues to a viable rescue situation. Once on the scene, responders need to evaluate the incident. Is rescue necessary? This information should be confirmed as best as it can be, but in some cases this information cannot be verified. What are conditions at the incident? Is there a confirmed chemical spill? If there is a release, can a person not wearing protective clothing survive? In many cases the local responders are called to suspected chemical releases and determines through investigation whether or not chemical release has occurred. If the persons who require rescue are alive, the actual risk to the reponders making the rescue is slight when wearing full PPE. If on the other hand the first responders arrive at a local mall and are told twenty people are unconscious and appear to be dead in the hardware store, the situation is different. When the responders approach the store entrance and see the twenty people lying on the floor not moving, the responders may not want to enter that environment.

Other considerations when making a rescue decision include response and notification time. If people have been in the hazardous environment for five to fifteen minutes, the response time of the department is critical to the decision-making process. The victims are not wearing any PPE and have been exposed to the material for a considerable amount of time. If they are alive when responders arrive, the risk is minimal to the rescuers. A method of emergency decontamination should be set up, and a backup crew should be standing by. The responding HAZMAT team should be consulted prior to entry to verify chemical infor-

mation. There should be no delay in evacuating the victims, and a swoop and scoop technique may be the best course. Stokes baskets or other methods of quick evacuation should be employed. Do not take vitals, ask medical histories, or perform medical procedures. Move patients to a safe area. Once out of the area, decontamination should be performed, if required for the chemical. The patients and rescuers require decontamination and isolation from the remainder of the responders. Once decontaminated, EMS personnel can begin to work on the patients using appropriate levels of PPE. After decontamination the rescue team should remove PPE and bag it. After evaluation by EMS providers and consultation with the HAZMAT team, the rescuers should be sent to rehab. Their PPE and SCBA will need to be evaluated and possibly sent for cleaning or replacement.

SAFETY The rescue of victims is made using simple risk/benefit analysis; a lot of risk is taken when a lot is at stake. No risk is taken when the benefit is little.

When handling hazardous materials incidents, encountering trapped victims is an unusual incident, but procedures should be in place to cover this contingency. The most likely scenario is a traffic accident in which persons are trapped and chemicals are involved. This scenario is common if we introduce the chemical gasoline to the picture. Firefighters are used to extricating trapped victims with gasoline leaking from one or more vehicles. Protection lines are established and typically a higher level of PPE may be used by rescue crews. The rescue crews limit the number of people in the hazard area, and a quick extrication is usually performed. If other chemicals are involved, the scenario may change slightly: as HAZMAT companies arrive, personnel can be switched out for those with better chemical protection, or the HAZMAT company can work to control or eliminate the chemical hazard.

To make the decision to enter a hazardous environment takes training and experience. This decision should be made in direct consultation with the HAZMAT company. To make that decision some HAZMAT teams are using risk-based response. This system is designed to identify the risk that chemicals present. By the heavy use of air monitors, an unknown chemical can be characterized into one of three risk categories: fire, corrosive, or toxic. When information is known about a chemical, it can still be placed into one or more of these categories by examining some chemical and physical properties. The major determining factor in a hazard that a chemical presents is vapor pressure. If the chemical does not have a vapor pressure, the risk of entering an environment with this spilled material is very low unless responders touch or fall into the material.

Another fact HAZMAT teams can use to their advantage is that the majority of the incidents to which they respond involve flammable and combustible liquids or gases. Although not designed as such, firefighters' TOG offers ample protection for a

rescue situation. HAZMAT teams and many first responders have great detection devices for these types of materials and can determine the true risk during an operation. **Table 5-3** lists the top ten chemicals spilled in this country every year. This is a HAZMAT team's bread and butter, and the team should be comfortable handling these materials. First responders should know the locations in which these materials are stored and used. The response to an incident involving these materials should be no different than a response to a bedroom fire.

SITE MANAGEMENT

Effective management at the site of a hazardous materials incident can alleviate the threat with a minimum of injury and property damage. Key management decisions are establishment of isolation zones and the need to evacuate or shelter citizens in place.

Establishment of Zones

Some jurisdictions may refer to zones as **sectors**. Whatever term is used, these areas are established to identify the various isolation points, such as those shown in **Figure 5-7**. These zones are referred to as

CASE STUDY

A truck carrying food rear-ended another tractor trailer stopped in the middle lane of the highway. The truck was traveling at an estimated 65 mph when it hit the stopped truck. The truck it hit was carrying a mixed load consisting of mostly 55-gallon drums. The driver of the stopped truck was not hurt and was able to remove his shipping papers.

The driver of the food truck was still alive but was extremely entangled in the wreckage, as can be seen in **Figures 5-4 and 5-5.** He was conscious and alert, and initial vital signs were stable.

The trucks were tangled together and the door to the truck carrying the drums was torn away from the truck. Drums had shifted onto the food truck and were laid over the front of that truck. The first responders saw the placards on the first truck and requested a HAZMAT assignment. When they approached the truck to evaluate the driver, the responders saw the drums in a precarious position in the back of the truck, as shown in **Figure 5-6.** They already had TOG on but also donned SCBA. They obtained the shipping papers from the driver and consulted with the HAZMAT company, which was about twenty minutes away. The contents of the drums contained mostly flammable liquids and

some combustible liquids. The first responders were advised to use a combustible gas indicator, continue with full PPE, establish foam lines, and begin the rescue.

Upon arrival the HAZMAT company members met with the IC to evaluate the scene. They confirmed the monitoring being done by the first responders and began to evaluate the other parts of the load. Whenever the rescue companies moved a part of the dash of the truck, one of the leaking drums would increase its flow. The HAZMAT companies secured the leak, moved the drum, and secured the remainder of the drums. They examined the rest of the load to make sure there were no other problems and continued to monitor the atmosphere. Once the victim was removed, the rescue companies moved away and the HAZMAT team **overpacked** and pumped the contents of the other drums into other containers. The risk category was fire, and the PPE chosen was appropriate for the risk category. At no time were any flammable readings indicated during the rescue, although some were encountered during the transfer operation. This was an example of the various disciplines working together to rescue a victim in a hazardous situation.

TABLE 5-3

Top Ten Chemicals Spilled
Sulfuric acid
Hydrochloric acid
Chlorine
Sulfur dioxide
Ammonia
Sodium hydroxide
Gasoline
Propane
Combustible liquids
Flammable liquids

These top ten (which are in random order) are derived from the EPA, OSHA, and DOT by examining the chemical release records published in a study as required by the 1990 Clean Air Act Amendments. As each agency tracks releases in a different fashion, we calculated each agency's top ten and determined an average; the result is this list. The category for flammable and combustible liquids is lengthy, and there was not enough information available to determine definitively a specific chemical for the final two entries.

Figure 5-5 Rescue crews are extricating the driver of a second vehicle involved in a rear-end collision. The driver remained conscious throughout the extrication and was severely entangled. Complicating the rescue was the presence of a truck full of hazardous materials containers. *(Photo courtesy of the Baltimore County Fire Department)*

Figure 5-6 Leaking drums were found during this rescue operation. Rescue crews had a foam line standing by and were air monitoring for flammable levels. When HAZMAT crews arrived, they took over air monitoring, quickly removed the leaking drums, and secured the other drums. When the driver was extricated, HAZMAT crews took care of the remaining drums. *(Photo courtesy of the Baltimore County Fire Department)*

Figure 5-4 The truck to the right rear-ended the front truck trapping the driver inn the truck on the right. In the trailer to the left are fifty five-gallon drums, some of which are leaking. Using PPE, air monitoring devices, and protection lines, crews continued with the rescue. HAZMAT crews evaluated the scene and handled and isolated the problem drums. *(Courtesy of Baltimore County Fire Department)*

SAFETY The most important zone that needs to be established is the **isolation area,** and this needs to be done by the first arriving responder.

the **hot, warm,** and **cold zones** and in many cases are identified by the use of colored barriers or cones. A comparison of the EPA designations is provided in **Table 5-4.**

Hot is usually identified by the color red, warm by the color yellow, and cold by the color green, as shown in **Figure 5-8.**

This area normally becomes the hot zone, after the arrival of the HAZMAT team. This isolation **hazard area,** or hot zone, is the area immediately around the release, which requires the use of PPE. Entering the hazard area exposes personnel to the

Figure 5-7 Proper isolation and work zones are needed to ensure the safety of responders.

TABLE 5-4

Hazard Area Designations	
Standard designation	**EPA designation**
Hot zone	Exclusion zone
Warm zone	Contamination reduction
Cold zone	Support zone

substance, and in some cases if contact is made, they would become contaminated. The minimum that should be established is this isolation area, and the distances for this area can be determined by the use of Emergency Response Guidebook (ERG). The minimum average distance established by this book is 330 feet, which should be the absolute minimum for an unknown material. The distance for this area is obviously determined by the local situation, and it is difficult to make blanket statements about recommended distances. If it is convenient and easy to isolate a block, then do so; if isolating the block may create havoc in the community, then 330 feet may be recommended. Keep in mind that if there are no indicators of an actual release, this is for an unknown material. Any indications of a flowing spill or a vapor cloud will dramatically increase this distance.

If time permits and resources are available, then responders should attempt to set up the warm and cold zones. These zones are usually established by the HAZMAT team after its arrival and setup. It is important that the first responders position themselves in an upwind and uphill position, as shown in **Figure 5-9.** At a water-based spill, responders should operate upwind and upstream. The first arriving apparatus should set up in that position, but if it does not, responders should move to that

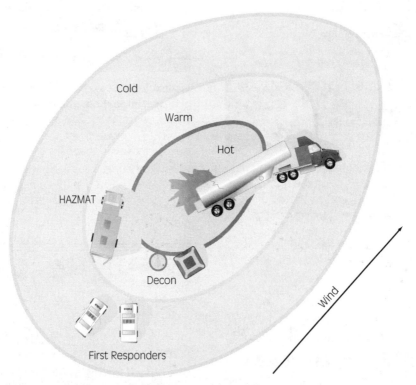

Figure 5-8 The establishment of zones is usually based on the types of hazards that may be presented. For general chemical spills, zones are hot, warm, and cold.

Figure 5-9 The best position for responders is uphill and upwind from the release.

position as quickly as possible. Although it cannot always be accomplished, being in an upwind position is the first priority and uphill is the second. If responders can only be upwind but are forced to set up in the downhill position, they should make sure to extend the distance. A common mistake is that the first arriving apparatus does not communicate the on-scene wind directions and the best route of travel to allow for the other apparatus to arrive and stage at the upwind and uphill position. All apparatus should be positioned so that they are pointed away from the incident so in the event of a catastrophic release or other emergency, the apparatus can be moved without turning around or backing.

Other considerations for the hot zone are topography, accessibility for responding units and other resources, weather conditions, and bodies of water, especially those for drinking water. Setup areas for the HAZMAT team should accommodate decontamination areas and dress-out locations. Other exposures must be examined, and part of the isolation process is limiting possible ignition sources. Public exposure potential must be considered, and this can be broken down into specific time frames such as short term (minutes and hours), medium term (days, weeks, and months), and long term (years and generations). Although emergency responders will usually only deal with short-term and medium-term incidents, the actions they take can result in long-term exposures for both the responders and the public. Sewer lines, both sanitary and storm, must be taken into account along with other utilities, such as cable and phone, both above ground and below

ground. Transportation corridors, such as highways, rail lines, ports, and airports, must be considered as well, as the incident may involve or affect these areas. The internal areas of buildings can be broken down into zones, but floor drains, ventilation ducts, and air returns must be taken into account.

The warm zone is set up after the arrival of the HAZMAT team and is usually where the decontamination area is established. In some cases the warm zone is also extended to allow for the HAZMAT team to set up. In unusual situations the warm zone is an area where some type of PPE may be required. Depending on the circumstances, it may be affected by a wind shift or catastrophic failure. The establishments of these zones can be arbitrary and are based on the best judgment of the IC or HAZMAT team. There is no magic line of hot and warm zones, nor could one be established. The cold zone is the area where the IC post is established and where the first responding companies will be positioned. All support operations such as medical and rehab are set up in this area, and movement between the zones is controlled at access control points. No PPE is required for the cold zone as there should be no chance for chemical exposure in this area. A person should be assigned to act as security for each of these zones to ensure that only authorized personnel enter these areas. Some HAZMAT teams employ monitoring to establish the zones and use the LEL, or exposure values, to determine these distances. The use of air monitors to establish these isolation points is crucial, as the distances provided in the DOT ERG are very conservative and are based on worst-case scenario. Real-time, on-the-scene air monitoring cannot be replaced by plume projections, estimations, or models depicted in a text. The distance that the zones are set up can be based upon the theories established by risk-based response. The materials that require the greatest isolation are those that have high vapor pressure or exist as gases in their natural state. A material such as sodium hydroxide (sometimes referred to as lye or caustic) only requires an isolation distance of a few feet. It is very corrosive and skin contact would cause some irritation and burns if not washed off immediately, but it has a very low vapor pressure. The vapor pressure of sodium hydroxide is 1 mmHg at 1390°C, which means that the spill would have to be heated to the 1390°C to produce even a small amount of vapor; that is considerably less than the vapor pressure of water (25 mmHg). With this material, it is unlikely that a large isolation distance would need to be established, nor would any evacuations probably be necessary. A chlorine release from a railcar, on the other hand, presents an extreme risk to the community: It may exist as a liquid in a railcar, but once in the atmosphere rapidly changes to a gas. Chlorine is not only poisonous but is an oxidizer and a corrosive material, and it has a very high vapor pressure of 5,168 mmHg. A substantial evacuation area is required for chlorine. One advantage is that chlorine is easily detected, and a true hazard area can be established using air monitoring.

Evacuations and Sheltering in Place

NOTE Making a decision to **evacuate** or to **shelter in place** can be a difficult decision for an emergency responder. No matter the decision, there are usually political ramifications, right, wrong, or indifferent.

The IC will be bombarded by the public, media, and coworkers about the evacuation decision. To a first responder, unfortunately, little assistance is readily available. The only resource is the conservative DOT ERG, which may not apply in your specific situation. The best way to determine evacuation or sheltering in place is real-time air monitoring that can establish the exact hazard area. Consultation with the HAZMAT team and the local conditions guided by the DOT ERG can establish a starting point. If the incident is at a SARA Title III facility, the jurisdiction's emergency plan should have some recommendations regarding evacuation and that facility. Some

Figure 5-10 Most HAZMAT teams use computer software to project the travel of a vapor cloud.

plans provide a checklist to follow to assist with the evacuation decision-making process. Close coordination with your local emergency management agency is essential to make the outcome successful. If you make the decision to evacuate, a suitable location needs to be found; transportation may be required and accommodations need to be established for evacuees. This type of assistance is usually provided by the emergency management coordinator, who is usually a good point of contact, and by the coordinating agency during an evacuation. The HAZMAT team can also run a plume projection, as shown in **Figure 5-10,** through the CAMEO computer program, but that is also conservative and may cause a greater evacuation than is necessary.

Chemical vapor plumes have characteristic shapes as determined by the computer models, such as the ALHOA model used by CAMEO. These standard plumes are to be used for worst-case scenario and may not apply to your locality. Although the plumes are computed for a variety of topography and weather, each local area differs, and only local weather conditions and air monitoring results can provide truly accurate results. The standard shapes for plumes are provided in **Figure 5-11.**

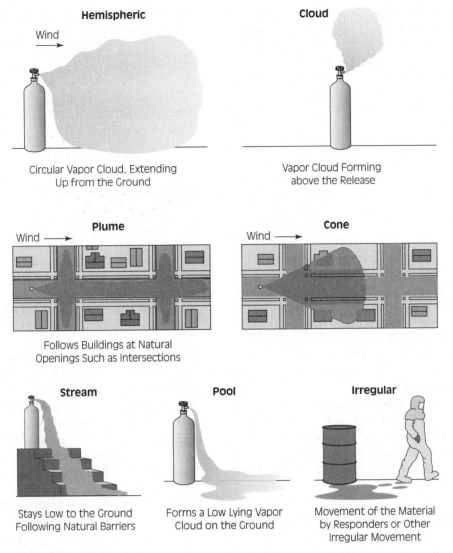

Figure 5-11 Standard plumes or vapor clouds may form after a gas is released. The exact type varies with topography and buildings in the area.

Some studies have determined that in most cases sheltering in place is safer than evacuation. Just imagine evacuating the most populated area of your community. How would you notify all citizens? In most communities the police department may have this responsibility, but do they have the resources to accomplish this task? Is there an emergency alerting system for TV or radio in your community? In urban or metropolitan areas a few blocks can require the evacuation of thousands of people. The worst-case scenario using the DOT ERG is an evacuation distance of 400 feet by 7 miles, which in a city could encompass 25,000 to 100,000 people in an evacuation. It would be nearly impossible to evacuate that many people, without widespread panic and chaos.

NOTE Typical evacuation situations can cause injuries and fatalities, even when the evacuation is announced days ahead of time.

Certainly, evacuation is required when an explosion is probable, explosives are involved, or the container may BLEVE or rupture with violent consequences. Evacuation may be the best option when the release will continue for a period of time, more than a few hours. However, an evacuation may not be recommended for a hospital, nursing home, jail, or other facility in which rapid removal is not practical. In these types of occupancies, air monitoring and control of HVAC are advisable; in addition, having a liaison remain at the facility who is in radio contact with the

IC is important. The chemical properties will have an effect on the decision, as some materials will rise into the atmosphere and dissipate quickly, while others may stay low to the ground, causing evacuation problems. The type of leak must be assessed. Can it be quickly and easily controlled, or is there a probability that it cannot be controlled by the HAZMAT team? Will evacuating the citizens subject them to a higher level of the chemical than keeping them in place?

The ACGIH has provided some planning levels for the chemicals that require planning under the Clean Air Act Amendments (CAAA). These planning levels are on three tiers and provide actual levels that can be used for emergency planning. As the permissible exposure limits and threshold limit values are established for an eight-hour exposure during normal conditions, they do not have much applicability during an emergency release. These **emergency response planning (ERP)** levels are designed to assist with the emergency planners' preparation of the community emergency plan and would be useful in determining the evacuation zone. When making the decision to evacuate or shelter in place, the flow of information to the public and the media is essential. A lack of information can bring disastrous results. When sheltering in place the citizens should shut all windows and doors, shut off air handling systems, and stay tuned to a TV or radio station. Without a continual information flow from the incident, citizens become frustrated and make attempts to find out the information for themselves. When dealing with larger facilities such as high rises, hospitals, nursing homes, schools, or jails, it may be best to put an emergency responder at that location to be a direct link to the incident, so that any questions can be immediately answered and fears alleviated. The emergency management office can also establish a rumor control hotline, which can be used to answer questions about the incident from concerned citizens and family members.

COMMON INCIDENTS

This section provides an overview of common incidents and the types of releases in each DOT hazard class. This listing is far from complete and focuses on the most common incidents or situations that have a great potential impact to the community. The recommendations provided here are only suggestions, and local policies and procedures should be followed.

Types of Releases

Chemical incidents can be classified based on how the chemicals are released from their container. In many cases the manner in which the release occurred pro-

vides clues to the successful mitigation of the incident. The type of release can be classified as a breach in a container or a release within a containment system. Although confusing, the NFPA provides two distinct categories for both breaches and releases.

In assessing chemical events there are several ways at looking at the potential release of a chemical from a container. Either the chemical itself is stressed or the container is stressed. In an incident involving a gasoline tanker that is rolled over at 55 mph the container is stressed, and the contents are likely to come out of the container. In an incident in which a paint waste is reacting within a drum, the material is stressed, but if the drum is sealed tight, the container will be stressed due to the chemical action creating pressure. There are three general types of stress: thermal stress, mechanical stress, and chemical stress. Thermal stress is the addition of heat or cold to a container; cold stress on a container can be as damaging as heat stress. Placing a metal drum into a pool of a liquefied gas would result in the drum becoming brittle and easily cracked. Dropping a drum is an example of mechanical stress. Putting a corrosive in a metal drum is an example of chemical stress. Both pressure and non-pressure containers can breach under a variety of conditions. **Figure 5-12** provides examples of each type of container breach. The most common container breaches are punctures and closures opening up. It is uncommon to find a container that is disintegrated or has runaway cracking, splits, or tears. Many factors are involved with container breaches, but the most common breaches are nail punctures in drums, forklift punctures, dropped drums, and closures not used or not in place. Many incidents have been quickly mitigated by the use of a drum lid. On tank trucks a common breach point is the frangible disk, which may rupture due to overfilling or a quick stop causing the liquid to slosh and rupture the disk, releasing some of the contents. The incident is quickly handled by the replacement of the disk.

The methods in which chemicals can be released through a **containment system** are detonation, violent rupture, rapid relief, and a spill or leak. When some of these methods release a chemical, the action can be violent and have catastrophic consequences. When a container such as a propane tank detonates, it can travel upwards of a mile, such as the one shown in **Figure 5-13.** If a container has a violent rupture, it means that the container was under pressure, causing the violent release. Much like a BLEVE a violent tank rupture can send the tank or portions thereof a considerable distance. A container that has a rapid relief hopefully releases enough pressure to bring a margin of safety to the incident. This is only true if the cause of the

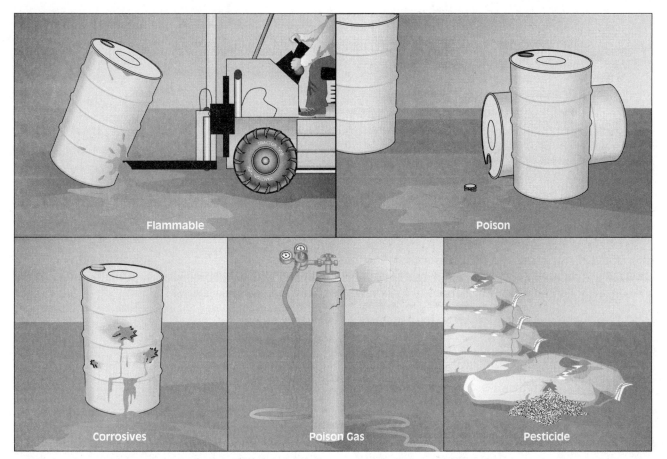

Figure 5-12 Examples of container breaches.

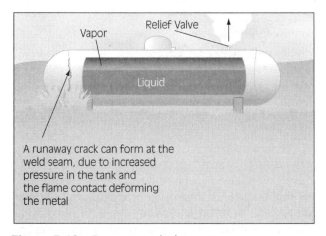

Figure 5-13 Propane tank detonation.

increased pressure is removed and the pressure does not continue to climb within the container. If a relief valve is operating, this constitutes a true emergency condition and failure to bring the pressure under control may mean catastrophic consequences. A relief valve operating is allowing the product to escape; if the material is flammable, it may find an ignition source and ignite. If the release ignites, this may cause the internal pressure to increase even more, causing more product to ignite. A relief valve

should never be stopped, nor a fire coming from the relief valve be extinguished. The only time to extinguish a fire coming from any valve (relief or otherwise) is if the flow of the material can be stopped with full assurance that it will be successful. Extinguishing any other fire, including those that are impinging on a tank or container, should be a high priority as should hose lines be directed to cool the container to prevent the relief valve from operating.

Spills and leaks are two events that result in loss of the product. A spill is defined as a loss of product from a naturally occurring opening, such as a **bung** on a drum or a leaking valve. A leak is defined as a release from an unintended opening, such as a puncture in a side of a drum. The result for a spill or a leak is that product is released and can create problems for responders.

Explosives

SAFETY The general rule with explosives is if the fire is near or is affecting the explosives, then emergency responders should isolate the area and back away. Life safety is the top priority.

RESIST RUSHING IN !
APPROACH INCIDENT FROM UPWIND
STAY CLEAR OF ALL SPILLS, VAPORS, FUMES AND SMOKE

HOW TO USE THIS GUIDEBOOK DURING AN INCIDENT INVOLVING DANGEROUS GOODS

ONE   IDENTIFY THE MATERIAL BY FINDING ANY **ONE** OF THE FOLLOWING:

   THE 4-DIGIT ID NUMBER ON A PLACARD OR ORANGE PANEL

   THE 4-DIGIT ID NUMBER (after UN/NA) ON A SHIPPING DOCUMENT OR PACKAGE

   THE NAME OF THE MATERIAL ON A SHIPPING DOCUMENT, PLACARD OR PACKAGE

   IF AN **ID NUMBER** OR THE **NAME OF THE MATERIAL** CANNOT BE FOUND, SKIP TO THE NOTE BELOW.

TWO   LOOK UP THE MATERIAL'S 3-DIGIT GUIDE NUMBER IN EITHER:

   THE ID NUMBER INDEX..(the yellow-bordered pages of the guidebook)

   THE NAME OF MATERIAL INDEX..(the blue-bordered pages of the guidebook)

   If the guide number is supplemented with the letter "P", it indicates that the material may undergo violent polymerization if subjected to heat or contamination.

   If the index entry is highlighted, **LOOK FOR THE ID NUMBER AND NAME OF THE MATERIAL** IN THE TABLE OF INITIAL ISOLATION AND PROTECTIVE ACTION DISTANCES (the green-bordered pages). If necessary, **BEGIN PROTECTIVE ACTIONS IMMEDIATELY** (see the section on Protective Actions).

   USE THE FOLLOWING GUIDES FOR ALL EXPLOSIVES:

   DIVISION 1.1 (EXPLOSIVES A) - GUIDE 112
   DIVISION 1.2 (EXPLOSIVES A & B) - GUIDE 112
   DIVISION 1.3 (EXPLOSIVES B) - GUIDE 112
   DIVISION 1.4 (EXPLOSIVES C) - GUIDE 114
   DIVISION 1.5 (BLASTING AGENTS) - GUIDE 112
   DIVISION 1.6 - GUIDE 112

THREE   TURN TO THE NUMBERED GUIDE (the orange-bordered pages) AND READ CAREFULLY.

NOTE   IF A NUMBERED GUIDE CANNOT BE OBTAINED BY FOLLOWING THE ABOVE STEPS, AND A PLACARD CAN BE SEEN, LOCATE THE PLACARD IN THE TABLE OF PLACARDS, THEN GO TO THE 3-DIGIT GUIDE SHOWN NEXT TO THE SAMPLE PLACARD.

IF A REFERENCE TO A GUIDE CANNOT BE FOUND AND THIS INCIDENT IS BELIEVED TO INVOLVE DANGEROUS GOODS, TURN TO **GUIDE 111** NOW, AND USE IT UNTIL ADDITIONAL INFORMATION BECOMES AVAILABLE. If the shipping document lists an emergency response telephone number, call that number. If the shipping document is not available, or no emergency response telephone number is listed, IMMEDIATELY CALL the appropriate **emergency response agency listed on the inside back cover of this guidebook.** Provide as much information as possible, such as the name of the carrier (trucking company or railroad) and vehicle number.

Figure 5-14   ERG includes some basic explosives information that provides a good margin of safety when dealing with these incidents.

All persons should be removed from the area, and a defensive operation should be established. A lot of other considerations come in to play if the fire is not directly affecting the explosives. A brake fire on a truck carrying explosives is one example. As long as the fire has not reached the cargo area, the fire may be extinguished without incident. When making the decision to fight a fire in this situation, water must be applied quickly and in large quantities. Crews should be limited, and adjoining areas evacuated. Water should be applied to cool the cargo and then extinguish the fire. The recommendations in the DOT ERG, as shown in **Figure 5-14,** are a good starting point for isolation and evacuation when involved in fire situations. Some recommendations provided by the federal Bureau of Alcohol, Tobacco, and Firearms (ATF) are provided in **Table 5-5.** Much like the ERG, these recommen-

dations are conservative and actual distances may vary. It is recommended that responders always consult with their local bomb technician for advice.

Other incidents that involve explosives include assisting the bomb squad with a suspected explosive device. Close coordination with the bomb squad is required to make sure that the operation is conducted safely. Although the distance varies with some jurisdictions, radios should not be used within 500 to 1,000 feet of a suspected device. Many bomb squads are detonating suspected devices in place, which may start a fire or cause a structural collapse. Other techniques are to "disrupt" the device or render it safe by the use of a water cannon. These techniques could cause the device to detonate. The time to discuss these scenarios and some plans of action is prior to the incident, not during the incident. At these types of incidents involving a suspected

TABLE 5-5

Bomb Threat Standoff Distances

Threat description	Explosive capacity	Lethal air blast range	Mandatory evacuation distance	Desired evacuation distance
Pipe bomb	5 lbs.	25 ft.	70 ft.	850 ft.
Briefcase or suitcase	50 lbs.	40 ft.	150 ft.	1,850 ft.
Compact sedan	220 lbs.	60 ft.	240 ft.	915 ft.
Sedan	500 lbs.	100 ft.	320 ft.	1,050 ft.
Van	1,000 lbs.	125 ft.	400 ft.	1,200 ft.
Moving van or delivery truck	4,000 lbs.	200 ft.	640 ft.	1,750 ft.
Semi-trailer	40,000 lbs.	450 ft.	1,400 ft.	3,500 ft.

Explosive capacity—based on the maximum volume or weight of explosives (TNT equivalent) that could reasonably be hidden in the package or vehicle.

Lethal air blast range—the minimum distance personnel in the open are expected to survive from blast effects. It is based on severe lung damage or fatal impact injury from body translation.

Mandatory evacuation distance—the range to which all buildings must be evacuated. From this range to the desired evacuation distance, personnel may remain in the building (with some risk) but should move to a safe area in the interior of the building away from windows and exterior walls. Evacuated personnel must move to the desired evacuation distance.

Desired evacuation distance—the range to which personnel in the open must be evacuated and the preferred range for building evacuation. This is the maximum range of the threat from flying shrapnel/debris or flying glass from window breakage.

Source: Developed by the ATF, with technical assistance from the U.S. Corps of Engineers. Supported by the Technical Support Working Group (TSWG), a research and development arm of the National Security Council Interagency working group.

device, be aware of the potential for a secondary device, designed to injure the responders. When operating at the scene of an explosion, it is important to have the bomb squad search the scene for secondary devices, as shown in **Figure 5-15.** Chapter 7 covers secondary devices in more depth.

Other incidents may involve a shipment of explosives that are involved in an accident; in most cases these incidents present little risk to responders or the community. In this case the bomb squad is a great technical resource and should be used to evaluate the scene prior to moving any explosives. In some cases an explosives truck, such as the one shown in **Figure 5-16** carrying ammonium nitrate, may be involved in an accident. The fuel oil on board the vehicle could mix with the ammonium nitrate, creating ANFO. Although without an initiation charge or other substantial energy source the material is unlikely to explode, the ammonium nitrate does present a toxicity hazard and care should be taken around spills of this material. In incidents of this type the bomb squad should also be consulted. Chapter 7

Figure 5-15 Handling explosives requires extensive training and equipment.

Figure 5-16 This truck has separate tanks containing ammonium nitrate and fuel oil. In the event of a crash, they may mix, but without an initiating charge the mix is unlikely to detonate. *(Courtesy of the Maryland Department of the Environment)*

discusses some possible terrorism scenarios, and explosives are used in almost all of the incidents. HAZMAT and bomb squad duties are becoming more and more interlaced.

SAFETY It is not uncommon to have explosives brought to the responders by well-meaning citizens. If the explosives are outside or in the citizen's vehicle, leave them there and call for the bomb squad.

These explosives may include hand grenades, pipe bombs, dynamite, blasting caps, ammunition, or distress flares. Responders should not handle these materials but should request that the bomb squad handle the situation. Depending on the type and size of device, isolation and evacuation may be necessary.

SAFETY It is a good general practice not to bring unknown materials into the responders work area.

Citizens may bring in old chemicals, some of which may be explosive and shock sensitive. With picric acid, for example, the simple removal of the cap or vibration of the container is enough force to initiate the explosion of the container, which may produce enough force to cause fatal injuries to the person who attempts to open or set down the container. Another common chemical that may be brought in are containers of old ether (ethyl ether), which is outdated after a year in storage. Opening this container may also bring fatal consequences due to the formation of peroxide crystals under the lid. These crystals can be shock sensitive and can detonate by the **shearing effect** when opening the container. The job of handling these containers usually falls to the bomb squad with a HAZMAT team interface. In some cases toxic materials may be brought in, which can contaminate the person who accepts the package and cause severe health problems for firefighters in the station. Depending on your state regulations, if you accept any package from a citizen, you may end up owning the package and may be required to pay for the proper disposal of the item. This can amount to a costly good deed.

Gases

Incidents involving gases include both flammable and nonflammable gases, with the most commonly released gases being flammable. Also on this category are poisonous gases; see the poisons section later in this chapter for more information on poisonous gases. Luckily, many responders are carrying gas detection devices that will warn of potentially explosive atmospheres involving flammable gases. Depending on the setup, the detector may warn of oxygen-deficient atmospheres, which may be caused by the release of a nonflammable gas.

The two most commonly released flammable gases are natural gas and propane. Common propane bottles are shown in **Figure 5-17**. Although used for the same purposes, natural gas and propane have differing characteristics that can affect a response to a gas leak. Both gases are odorless when they exist in their natural state; when moved through a distribution system, an odorant is added to them. In major interstate pipelines it is possible that the gas may be traveling without an odorant added.

Figure 5-17 These 20-pound cylinders, which many homeowners have, can create large fireballs and explode with considerable force if involved in fire.

NOTE The largest difference between natural gas and propane is the vapor density. Propane is heavier than air and will stay low to the ground trying to find an ignition source. Natural gas, on the other hand, will rise in air and should dissipate quickly.

This dissipation is weather dependent, and responders may encounter weather conditions that affect these characteristics. Under some conditions the gas may travel the length of a pipe following the path of least resistance, usually coming up inside a building. The leak may not be detected for a few days or weeks, until it surfaces. This is most common in areas of the country where the ground freezes and limits the upward movement of the gas. In many states specific training is offered on natural gas and propane emergencies, in most cases provided by gas companies.

Incidents involving these gases are commonplace. For natural gas the most common incident involves a ruptured gas main. Much like a municipal water system that is set up in a grid system, the gas system is set up on a grid, with larger distribution pipes and smaller delivery pipes. The pipes leading into a home are typically ½ inch to 1 inch in diameter. Distribution pipes may be 2 inches up to 36 inches depending on the region. First responder actions at these types of incidents generally involve isolation and protection, followed by standby until the line can be shut off. Gas detection devices should employed to determine the true hazard area, and

adjacent buildings should be checked for gas. When checking buildings, air monitoring should done at a variety of points, including the highest point, which should be checked prior to declaring a building safe.

CAUTION At no point should first responders jump into a hole to attempt to shut off a leak, nor should underground valves be shut in an attempt to stop the leak.

Gas line valves may actually be keeping the amount of gas reduced; moving the valve may increase the amount of gas escaping. If the leak is on the outtake site of the meter, it is acceptable to shut off the valve on the meter. The meter should be locked out and tagged out, and only the gas company should turn the flow of gas back on. Some departments carry special tags that mark the system as being out of service, and some actually lock the system in the off position. When gas company employees arrive, they repair the leak and place the system back in service. When dealing with pipes and electrical systems in an industrial setting, an OSHA regulation prescribes the procedure for shutting off a system and marking it so that the system is not accidentally turned back on prematurely. This regulation is known as the *lock out, tag out* regulation and applies to many emergency service situations.

Propane releases generally involve cylinders, although it is possible to have a pipeline propane leak. A number of pipelines running across many states carry both natural gas and propane. Other than flammability, the fact that propane sinks and stays low to the ground increases the hazard. If the cylinder is releasing liquid propane, then the liquid is also becoming very cold, below 0°F. A common incident involving propane cylinders, especially those 20-pound cylinders designed for home barbecue grills, is overfilling the cylinder. Cylinders are designed to be filled to 80 percent of capacity to allow for expansion of the gas. If the cylinder is filled more than 80 percent and the temperature increases, the gas may escape the relief valve.

For fires involving propane tanks, there is the potential for a BLEVE, an event that can be catastrophic to responders. A BLEVE can occur with any flammable gas storage tank but is usually associated with propane tanks. If fire impinges on the tank and increases the temperature within the tank, the pressure also increases. As the temperature increases toward the boiling point, even more vapors are being produced and the pressure is increasing. At a certain point the pressure increases enough to cause the relief valve to operate. When this relief valve oper-

(A)   (B)

Figure 5-18 Vehicles that use alternative fuels such as natural gas, propane, or electric are marked to indicate the fuel source. Photo A shows a CNG sticker on the rear bumper, indicating that the car is fueled by compressed natural gas. The sticker in photo B is beside the front driver's quarter panel and indicates this vehicle is environmentally friendly alternatively fueled. Cars with alternative fuels typically still have a gasoline tank, and there may be gasoline present in emergencies.

ates, the pressure in the tank will force gas from the tank and in most cases cause an ignition of the releasing gases. Under these circumstances this is good for responders as gases that have not ignited can cause more problems. If the pressure in the tank is not reduced, the potential is that the pressure inside the tank will increase and overwhelm the relief valve and the tank will fail. When pressurized tanks fail, they can travel for a considerable distance. Although no one can predict exactly when a tank will fail, some clues may indicate that a BLEVE is forthcoming: increased flame height or appearance that the flames coming from the relief valve are under a higher pressure. The sound of a relief valve is deafening, but when the pressure increases, the pitch of the sound will get higher and may become louder.

SAFETY If the tank becomes discolored or distended or loses shape, then an immediate withdrawal is indicated.

When making the decision to fight a propane tank fire, a large quantity of water needs to be applied quickly and continuously. Responders should concentrate water flow on the vapor space of the tank, and the fire should not be extinguished unless responders are certain the flow of gas can be stopped. When preplanning, first responders should establish a flow of water on the tank in excess of 500 gpm within a few minutes of arrival. BLEVEs are respon-

sible for a number of responder deaths each year, and in some cases whole alarm assignments have been killed and seriously injured during BLEVEs. A container does not need to have a flammable liquid in it to BLEVE. Any heated liquid produces vapors that will cause the pressure inside a tank to increase and may cause the container to fail. This would be called a violent tank rupture (VTR) and would not have the accompanied fireball associated with a BLEVE. A drum of water that is heated can create a VTR and can be very dangerous to people nearby.

Some cars now are powered by natural gas or propane; the most common are cars that are part of a fleet of vehicles, including government and utility vehicles. To determine if a vehicle is powered by one of these gases, look for a sticker that reads CNG, LP, propane powered, or natural gas powered. Sample stickers are shown in **Figure 5-18.**

Other common gas releases involve carbon dioxide, chlorine, and ammonia. Although carbon dioxide (CO_2) is a common fire extinguishing agent, it is also used for the distribution of beverages and is found commonly in restaurants, bars, and convenience stores. An increase of CO_2 in a building can make people sick and is becoming a common **sick building chemical.** Although carbon dioxide is not commonly monitored, when dealing with emergency response to a **sick building,** responders should determine CO_2 levels. Both chlorine and ammonia have good warning properties and have distinct odors and should be easily identified by people in a building. In most cases the odor is so irritating that people self-evacuate the building prior to hazardous levels being built up.

Figure 5-19 Responding to this emergency presented several challenges, due mainly to the final resting place of the overturned truck. Although HAZMAT response prefers an uphill position, the best place for the offload truck was downhill. Extra distance and additional protection lines were necessary to compensate for the disadvantaged position. (Courtesy of the Maryland Department of the Environment)

Flammable and Combustible Liquids

By far, this is the leading category for the most common type of releases, as it includes gasoline and diesel fuel, as shown in **Figure 5-19**. Responders across the country respond to these types of incidents thousands of times a day.

SAFETY Gasoline is the leading chemical when it comes to chemical accident fatalities, and transportation leads in those fatalities. Unfortunately, due to gasoline's familiarity many emergency responders do not adequately protect themselves when responding to incidents of this type.

In some cases, responders themselves may improperly use the chemicals in the station,

unaware of the hazards. Gasoline has between 1 percent and 5 percent benzene, which is a confirmed carcinogen to which responders are commonly exposed. When you smell gasoline, you are receiving an exposure to benzene, an exposure which is repeated on a regular basis. When dealing with small spills, the tendency is to use little or no protective clothing, but the body is continually being exposed to this toxic material. In addition to its toxicity, gasoline is also very flammable and can be ignited by the static electricity generated by responders' clothing as they approach an incident. Although a lesser concern with diesel fuel, fuel oil, or kerosene, toxicity remains a hazard. In many cases, a lower level of concern has resulted in responder injuries. Although in normal circumstances it is difficult to ignite these materials, if the temperature is increased, they can be easily ignited. On a day with a temperature below 70°F, the potential for ignition is low, but a spill onto blacktop when it is 100°F outside adds considerable risk for a fire as a larger quantity of vapor will be produced on the pavement, which is likely in excess of 110°F. In years past it was commonplace to just flush spills of flammable or combustible liquids down the storm drain. This practice for the most part has been discontinued, as it causes severe environment damage and only moves the problem to another location. Responders in many areas of the country absorb all the material and dispose of it in an environmentally sound manner. To provide the best protection for the environment the material must be picked up. Discharging oil (includes fuels) is a violation of the federal Clean Water Act, and emergency responders could face serious fines if the materials were flushed into a waterway.[2]

Flammable Solids/Water Reactives/Spontaneously Combustible

When dealing with materials in these categories, specific identity and emergency response information is crucial. Consultation with the HAZMAT team is important, as a wrong tactic can be devastating to the community. Most emergency responders have some experience with flammable solids, as road flares are classified as flammable solids. In most cases flammable solids are difficult to ignite, but once ignited they burn vigorously and are difficult to extinguish.

The water reactives group can be defined two ways. When some chemicals get wet, a violent reaction such as a fire or an explosion may result. This is commonly what most people think of as water reactive. The sec-

TO FIGHT OR NOT TO FIGHT?

A common situation involves overturned or burning tanker trucks. The most common product in these trucks is gasoline, and when one of these trucks is involved in an accident, a fire is usually the outcome. Two examples of gasoline tanker fires are depicted in **Figure 5-20.** When first responding companies arrive, they have a tendency to try to extinguish the fire. Without larger quantities of firefighting foam, it is very difficult to extinguish a tanker fire. If upon arrival, the truck is well involved and is not near any exposures or impacting an adjacent community, the best course is usually to let the fire continue to burn.

When attempting to extinguish this type of fire, the considerable runoff usually contains both gasoline and firefighting foam. Both of these chemicals are not good substances to allow to enter a waterway. If the fire is extinguished, a large amount of gasoline at an elevated temperature may remain. Once extinguished, the truck must have foam reapplied every few minutes to ensure that the fire does not reignite, as shown in **Figure 5-21.** It is then necessary to remove the hot gasoline from the truck and pump it into another truck, a dangerous proposition. If, on the other hand, the fire is allowed to burn, nothing would remain that could harm responders or the environment. Although it looks horrible, the resulting black smoke is in reality less damaging to the environment than a liquid spill. The smoke is predominately carbon, which provides the thick black smoke.

If the fire is threatening a community or adjacent structure, then all attempts should be taken to extinguish the fire. In many cases the truck is wrecked under a bridge, which should be considered a critical structure in the community. The cost of a bridge usually is in the millions, and responders should assess the financial impact in terms of hard dollars and inconvenience if it were closed for a year for reconstruction. If units arrive and are confronted with a large fire, the initial water should be directed toward the bridge. When sufficient foam and water arrives, firefighters should then extinguish the truck and continually apply foam. The application of foam at a nonfire incident usually causes friction between firefighters and environmental agencies. Foam can damage the environment, and even if "environmentally safe" foam is used, the breakdown products of the fuel are not environmentally safe. Environmental agencies would prefer the minimal use of foam. Another concern is that when a truck is overturned, it must be drilled and pumped out prior to righting the truck; the application of foam makes the situation very slippery and creates additional hazards. Foam should only be applied on the gasoline, and coating the top of the truck is not necessary when the gasoline is on the ground. The use of air monitors to determine when the foam blanket is breaking down is recommended, as is the application of the least amount of foam to provide a foam blanket. An example of other hazards that may exist at a tanker rollover is described in **Figure 5-22.**

(A)

(B)

Figure 5-20 A & B These photos provide a graphic representation of the remnants of a gasoline tanker fire. In photo A all of the gasoline was consumed in the fire and the environmental damage was very slight. The tanker wrecked in an area that was not impacted by the fire, and it could be allowed to burn. In photo B this tanker had to be extinguished, because it slammed into and under a major bridge. The majority of the tank was intact, and in most cases, only parts of the tank melted down, exposing the gasoline inside. Out of the 9,000 gallons it was carrying, between 3,500 and 4,000 gallons were left on the truck. This created a very dangerous situation, especially during the transfer operation. The reapplication of foam on a regular basis in this case was essential; the truck reignited after it had been extinguished. *(Courtesy of the Maryland Department of the Environment)*

Figure 5-21 Although a necessary evil when dealing with flammable liquids, responders should minimize the amount of foam applied at a spill. It can cause further environmental damage and slip, trip, and fall hazards.

Figure 5-22 Not all hazards come directly from vehicles or chemicals when responding to highway accidents. This diesel fuel truck was offloaded safely, but the next day all of the response crews came down with poison ivy; some cases were very severe. *(Courtesy of the Maryland Department of the Environment)*

ond method is that once a reactive metal is ignited, the application of water may create a violent reaction. Magnesium is one example of this type of reaction. Once ignited, it burns vigorously, and when water is applied, it may explode, as shown in **Figure 5-23.** Some materials such as calcium carbide are shipped as water reactive materials. When water is applied to calcium carbide, the reaction may amount to a little bub-

bling, something that would not be considered violent. However, the gas that is being released by the bubbles is acetylene gas. When pure acetylene gas is produced in this manner, it is unstable and reactive. Any nearby ignition source will produce a fire or explosion. Thus, the actual application of water in itself does not produce a significant problem, but the action of the water may create additional concerns.

(A)

(B)

Figure 5-23 A & B Photo A shows a fire in which a measuring cup of magnesium shavings was added to burning straw and diesel fuel. The fire looks normal and does not indicate any unusual materials. Photo B shows what happens when a slight water spray is added over the fire. There is a dramatic increase in the size of the fire, and the resulting detonation spreads white hot particles of metal in the pan.

Materials that are spontaneously combustible are usually transported in a manner to keep them stable. To remain stable they may require specific temperatures or have a material added to them to keep them stable. White phosphorus, which is usually covered with water, is an example of a spontaneously combustible material, as it ignites in the presence of air. If your community has a facility that uses flammable metals, it is a good idea to have a stockpile of metal extinguishing agent such as metal-x or lith-x, commonly referred to as class-D extinguishing powder. Most HAZMAT teams have a limited supply of this type of extinguishing agent and usually rely on a local industrial resource for more. Sand that is known to be dry (it has been stored inside in a closed container) can also be used, but most sand contains moisture.

SAFETY The use of water on flammable metals is not recommended and can cause severe injuries to the hose crew.

The metal has a tendency to explode when water is applied, sending hot metal and other fragments flying and producing other hazardous chemicals. The water will break down and release hydrogen gas, which will increase the amount of heat and flame produced. When sodium is put into water it will detonate violently, and it also makes sodium hydroxide, which is a very corrosive material.

Oxidizers and Organic Peroxides

This is another class of materials for which responders should request help early in the incident. Both of these types of materials can have explosive characteristics and can lead to fatal mistakes. Organic peroxides can react explosively even if the responders have taken no action. A well-known oxidizer is ammonium nitrate. By itself it is relatively harmless, although it does pose some toxicity. It cannot explode unless it is mixed with a fuel in the presence of an initiating charge. Although all three items should not be transported together, there may be occasions that they might be on the same truck. In other situations the oxidizer does not require any oxygen to produce a fire, as it will provide its own. Oxygen is labeled with a specific oxygen placard or can be shipped with an oxidizer placard. Oxygen itself cannot be ignited, but it will greatly intensify a fire when it is in involved in amounts greater than the 20.9 percent, the typical atmospheric level. Liquefied oxygen (LOX), which is a cryogenic, presents

Figure 5-24 A leak of liquid oxygen on asphalt can present a shock-sensitivity problem and increase the risk of a fire.

even more hazards in addition to supporting combustion. On asphalt or another hydrocarbon, LOX makes the surface shock sensitive so that it can detonate if compressed, such as by a firefighter walking on it. Another concern is the absorption of oxygen in TOG from a LOX release, which can present a flammability problem. Caution should be used and the gear should be given time to air out, but this is a very rare occurrence. Most hospitals have large upright cryogenic LOX storage tanks, as do many nursing homes. An example is shown in **Figure 5-24.** Smaller in-home versions are available as well, presenting a large fire risk in a residential home, and a freezing and contact hazard if knocked over.

Another commonplace situation involving oxidizers is incidents caused by pool chemicals. Most of these chemicals are oxidizers or may react with oxidizers and can be involved in a chemical reaction or fire. If the materials do not ignite, a dangerous situation can develop, as a large chlorine vapor cloud can be created from just a handful of pool chemicals, which can produce disastrous effects on a neighborhood. Specific advice from the HAZMAT team is required prior to handling or disposing of any of these types of chemicals. There is a lessened level of concern with these types of chemicals because they are considered household products, but they do present a large hazard to the community.

Poisons

Class 6 refers to poison liquids, and we will include poisonous gases from class 2 (gases). When dealing with these types of materials, including "stow away from food stuffs" and "marine pollutant" materials, keep in mind that they are toxic to both humans and the environment. Materials labeled as poisonous should be considered very toxic, as the DOT reserves this class for highly hazardous materials. The predominate hazard a chemical poses places it into one or another category. There has been some improvement with the subsidiary placarding and labeling category, but there is still room for more.

Poisonous gases may also be accompanied with a label on a bulk container "poisonous by inhalation," which would be listed on the shipped papers as PIH. By the definition these materials present a risk to humans and the community and require extra precautions. The most common incidents with these types of materials are from pesticides and agricultural chemicals. Although most household materials are of lesser concentrations, there are occasions in which higher strength materials have been used. In several incidents, insect spraying companies (or individuals) have used full strength agricultural products in residential homes, requiring evacuations lasting a few weeks and removal of some homes due to the contamination.

SAFETY Technical grade pesticides are very dangerous and should be handled with care.

Incidents involving commercial home pesticides and fertilizing trucks are common, but in most cases this material is diluted and ready for home application. The fertilizers carried on these vehicles do not present much risk to the responders or the environment unless present in large quantities. The pesticide or insecticides do, however, present a risk to the responders and the environment. Responders should consult with the HAZMAT team prior to taking any action. Many home pest control vehicles may carry a mixed load of premixed and undiluted materials. When these vehicles are involved in collisions, responders should use full gear and anyone who comes into contact with any liquid or solid material should be decontaminated.

Radioactive

Although considerable emphasis is placed on radiation in the early HAZMAT classes, due to a low number of incidents involving these materials, some of the emphasis has been redirected to other areas. It is true that incidents involving these types of materials are rare, but they do have a significant impact on the well-being of the community. With the increased threat of terrorism, radiation has now been given more emphasis. Radioactive materials are commonly used in the community in smoke detectors, ground imaging equipment, and medical facilities. Although there is considerable concern when around radioactive materials, most of the materials responders would come across would not be harmful even for a lengthy exposure.

SAFETY Keeping radioactive material in the container intended for such use is paramount to reducing the exposure levels of the responders.

Some radioactive materials, such as those coming to and from a nuclear power generating facility, can present some risk to the community. Radioactive materials that present a risk to the community are shipped in high-strength containers designed to withstand a substantial crash or a fire situation without any release of radioactive materials. These shipments are usually well tracked and when involved in an incident, specialized help is usually already on the way before responders call for it. Knowing the local contact for radiation emergencies is key to an effective response when dealing with radioactive materials.

Corrosives

Behind the flammable and combustible categories, the next most common incident will probably involve a corrosive. Common corrosives are sulfuric acid, hydrochloric (muriatic) acid, and sodium hydroxide. Both of the acids are transported as liquids, but the sodium hydroxide may be transported wet or dry. Both sulfuric acid and sodium hydroxide have little or no vapor pressure and present little risk outside of the immediate spill area. Hydrochloric acid, on the other hand, has a high vapor pressure and may require some additional isolation and evacuation. To handle any of these materials, chemical protective clothing is required; after some time they will eat a hole in TOG. A common misconception is that spilling sulfuric acid on your hand will make your fingers fall off.

WEIRD INCIDENTS

Not all chemical investigations involve toxic chemicals, but until the HAZMAT team can determine the identity of the product, first responders should isolate the area and deny entry.

One HAZMAT team was called to assist an engine company at a marine terminal. The engine company reported a sea box placarded flammable was leaking an unknown colorless liquid. Upon arrival the engine company isolated the area, established the hot zone, obtained the shipping papers from the company, and began the research process. Nothing on the shipping papers seemed to fit the leak. The HAZMAT team chemist, who was also on scene, was equally confused. The HAZMAT team suited up and proceeded to investigate the truck with air monitors. At the back of the truck was a clear colorless liquid running from the corner of the truck onto the pavement. The liquid seemed to be evaporating and leaving a crystal-like material on the ground, as shown in **Figure 5-25.** None of the air monitors indicated any problems. The HAZMAT responders opened the back of the truck and revealed several 55-gallon drums. One drum was tipped on its side, and was found to be leaking **(Figure 5-26).** The HAZMAT team entered the truck to check the label and spelled out, "H, O, G, C, A, S, I, N, G, S," which produced a confused look on everyone's face. This hog casings (intestines) incident took a few minutes to wrap up.

About a month later, again at the marine terminal, the same HAZMAT was sent to investigate a ship that was having a "problem" with some containers. When responders arrived, they found that a ship had lost twenty-three sea containers at sea and now had seventeen containers leaning at an extreme angle. All but one of the containers contained food products; the one of concern was a container of phosphorus pentachloride, a very reactive material. From information obtained through the Coast Guard, Maryland Department of Environment, and the ship's crew, the HAZMAT team determined that the container of chemicals was involved but not damaged. The team boarded the ship to get a better view of the containers, which are shown in **Figure 5-27.** On the deck was an unusual odor, one that was faintly familiar. The responders all looked at each other with a quizzical look, trying to determine what the odor was; they were concerned that they might have been misled as to the safety of the ship. One of the bottom containers was crushed and the door was open. Responders looked inside to find the source of the odor—more hog casings. In fact, the sixteen other containers held more hog casings, with one HAZMAT container in the middle. Dealing with the HAZMAT container became a serious issue after it was determined how to offload the other containers, but it was interesting that hog casings, which are used to make sausages, were involved in another marine terminal event.

Figure 5-25 This photo shows the resulting leak from a sea container. In the middle is solid material that was left as the liquid evaporated. *(Courtesy of the Maryland Department of the Environment)*

Although this is an exaggeration, any splash of these materials should be washed off as soon as possible; actual burns will take a couple of minutes to take place and will progress in severity if the chemical is not quickly washed off. This applies for these materials only. Some uncommon acids, such as oleum or sulfur trioxide, will cause immediate burns upon contact and disfiguring injuries over a period of time. Chemical neutralization may be the best choice to handle a corrosive spill. This operation needs to be performed by a HAZMAT technician who has donned appropriate levels of PPE. When neutralizing, heat and violent reactions may be possible, so this action should be taken only after consultation with a chemist. Although one would think that simple dilution would take care of an acid spill, this is not true. A 100-gallon acid spill may require thousands, if not hundred of thousands, of gallons of water to bring the spill back to a neutral pH level.

Figure 5-26 This photo shows the back of a sea container, with a leaking drum on the left. One never knows what conditions will be found when opening the back of a sea container. In this case there was an upside-down drum and a knocked-over washing machine. More "junk" was found further up in the front of the container mixed in with the hazardous materials and hog casings. In some instances wooden braces at the back of the sea box require the use of tools to remove. The easiest removal method may be a chain saw, as long as flammability is not a concern. *(Courtesy of the Maryland Department of the Environment)*

Figure 5-27 On this container ship these 17 containers were listing severely; only one container in the middle contained hazardous materials. It was a difficult operation to remove the containers without toppling the entire pile and damaging the HAZMAT container. *(Courtesy of the Baltimore County Fire Department)*

Other Incidents

It is impossible to outline each specific action first responders should take at a chemical release. The first rule if responders suspect that hazardous materials are involved is to isolate the area and evacuate people in the immediate area until further information is available. Request the assistance of the closest HAZMAT team as its members will be needed for their technical expertise and equipment. First responders are not trained or equipped to provide detection or characterization of unknown materials. Depending on senses such as smell is playing Russian roulette.

SAFETY The absence of a vapor cloud or odor does not mean a potentially deadly material is not present.

Many toxic materials are odorless and colorless and provide little warning prior to causing lethal effects. It is not the job of first responders to seek out the cause of a chemical release nor to provide an identification. If the information is presented to them or can be obtained without risk, then it is of benefit, but otherwise it is the HAZMAT team's job to perform those tasks.

Common incidents that first responders may get involved with are odor complaints that may be classified as sick buildings, such as the one shown in **Figure 5-28.** The use of the HAZMAT air monitors is key to survival with these types of incidents. Always remember that your nose is not sufficient to determine immediately dangerous levels. When involved in sick buildings or odor complaints, the best course of action for first responders is to determine the validity of the complaint. First responders should interview the people involved and assess the likelihood that they are in fact experiencing the problems they state. If a number of people who work in a variety of locations throughout the building are complaining of similar signs and symptoms and report an identical odor, then the complaint is probably valid. Once it

Figure 5-28 A common incident at a high rise is a sick building in which the occupants become symptomatic from a variety of potential sources.

is determined that the call is valid and not related to a sunny Friday afternoon, first responders should don SCBA and request the services of the HAZMAT team. The occupants of the building should be removed, and the windows and doors should be shut. The HVAC system should be shut down so as not to remove the unknown material. These steps are crucial in the success of the HAZMAT team finding the source of the problem.

First responders may also get called to gas leaks inside a building, and their ultimate safety hinges on air monitoring, which is further explained in chapter 6. It is impossible to determine a level of gas in a building based on smell. Once in an environment that has a gas odor, the body desensitizes itself to the odor, and responders will not be able to smell the gas anymore, even if the amount has been increasing and moving closer to a potential explosion.

Gas grills are prone to propane leaks, and the leak can reach an ignition source, causing severe damage to the adjacent buildings. Many HAZMAT teams carry pipe flares to burn off the gas and prevent an explosion. First responder actions should be limited to isolation and evacuation and preparing standby hose lines. If the vapors are traveling toward other homes, then the use of hose lines with fog nozzles to move the vapors is appropriate.

DECONTAMINATION

The task of **decontamination** in some cases may fall to a person trained at the first responder operations level. If a first responder is expected to perform decontamination of a HAZMAT team, specific training in that area is required along with training in the use of chemical PPE. Decontamination is the physical removal of contaminants from people, equipment, and the environment. It is important to note that the concepts provided here may be for one or all three of these areas of concern. If one type of decontaminating solution is effective for a tool, it does not mean that it can be used on humans. Prior to performing any decontamination procedure on humans, responders must consult with the HAZMAT team or a chemist. Some decon solutions use very dangerous materials, especially upon contact with skin. For equipment it may be acceptable to use large quantities of sodium hydroxide mixed with detergent, but this same solution would cause severe burns if placed on the skin. When using decon solutions on humans, caution and consultation are required. There are two sources of contamination: direct contact or secondary contamination. Going into a spill area and putting your hand into a drum of blue paint is an example of direct contamination. If you leave the area and shake hands with your partner, this is an example of secondarily contaminating your partner with the blue paint.

Types of Decontamination

The four general types of decontamination levels are: **gross decon, formal decon, fine decon,** and **emergency decon.** In a full-scale incident, all four may be used, but in reality the majority of the incidents only require a minimal decon setup. Within each of the two large categories of decontamination, wet and dry decontamination, are four levels with some modifications. The two categories should be self-explanatory: one requires the use of water or other liquid solutions, while the other does not involve a wet process. The process of decontamination is chemical specific, and there are no absolute rules to the process. First responders should as a minimum have a good understanding of emergency decon procedures and develop a local procedure to handle this type of problem. Being confronted with a contaminated patient who is in need of decon is not the time to develop the procedure. With potential terrorism in mind, not only should responders think about the emergency decon of one person but of decontaminating thousands of people.

Figure 5-29 One of the simplest forms of emergency decon is the use of a hoseline.

Emergency Decontamination

Emergency decon can be as simple as washing someone off with a hoseline, though it can be further complicated by involving an unconscious patient on a backboard or stokes basket. For most chemicals, plain water washing is sufficient to decontaminate the person once clothes are removed. By removing the patient's clothes, you can eliminate 60 percent to 90 percent of the contamination. Military research involving warfare agents, even toxic nerve agents, shows that if the victim has been exposed to an aerosol, just having the person stay in fresh air for fifteen minutes can be an effective decon method. It would always be recommended that a person in this situation be decontaminated with a soap and water solution.

The common methods of emergency decon involve the use of a water line and some method of runoff control, such as shown in **Figure 5-29.**

> **NOTE** When people's lives are on the line, runoff is not a primary concern, but when not under those conditions, attempts should be made to recover any runoff.

One way to contain the runoff is to use an inflatable children's pool that can easily be carried on a fire truck. Another method involves the use of a tarp or plastic that is spread between two ladders. If none of these methods is possible, responders should use a tarp or plastic to catch the runoff as best as possible. If all other methods are not available, responders can seek natural areas that would collect the runoff. This type of decon should not be performed over storm drains or near areas of environmental concern. An industrial plant may have safety showers, and it is acceptable to use these for decon. Prior to using the showers, ask the people from the facility where the runoff goes. In most cases the runoff will go to the facility's own water treatment area and can be processed by that system. Caution should be used when the safety shower drains straight to the sanitary system; prior to using that shower, contact the local sewer authority.

Another general rule with regards to emergency decon applies to the removal of corrosives from a victim. Large quantities of water should be used. When the water mixes with the corrosive, there is the potential for a heat increase, and massive amounts of water will eliminate any potential for further injury. Besides using large quantities of water on corrosives, it is necessary to continue the flushing for a minimum of twenty minutes. No matter how bad the temptation is for EMS providers to remove the patient, the best treatment for the burns is the water flush, especially for burns caused by a base, such as sodium hydroxide.

Gross Decontamination

The process of gross decon entails the removal of the majority of the contaminants. It is usually the first rinsing station within a full decon setup, as shown in **Figure 5-30.** In many cases gross decon is just part of the whole process, but it can be the only step in a minor setup. Some HAZMAT teams use shower setups for gross decon, while others may use hoselines or pump tanks. As this is the most likely location for contaminants, attempts should be made to recover any runoff. When part of a multistep process entry team members usually perform this step on themselves without any assistance. When dry decon is used, there may be occasions that overgarments such as booties or extra gloves are used. When leaving the hot sector, there is a container or location to put dirty boots and gloves. For many materials dry decon is an acceptable use of PPE and resources. When using any type of decon, the bottom line is to make it safe for the responder to get out of the PPE and to create a safe situation for the responders assisting with PPE removal. When using reusable PPE, decon needs to be more formal, but even extensive cleaning can be done to the PPE after the incident and removal of the responder.

Formal Decontamination

The process of formal decon involves actual scrubbing and cleaning to remove any residual

Figure 5-30 Gross decontamination is used to remove the majority of contamination.

Figure 5-31 Formal decontamination is used to remove any further contamination that may remain after gross decon.

contaminants, as shown in **Figure 5-31.** It is usually done after the gross decon, typically in a couple steps. Attempts should be made to recover any runoff, but this is not as critical as with gross decon. During this procedure decon solutions are typically added to the process. These solutions are chemical specific. This process may involve showers, hoselines, or pump sprayers, and usually one to two people dressed in a lower level of PPE perform this activity. During formal decon it is important to pay particular attention to the areas most likely to be contaminated: hands and feet.

Fine Decontamination

This form of decon is not performed by first responders, but by hospital-based personnel. This cleaning removes all contaminants from the body. It would be nearly impossible to clean eyes, ears, fingernails, and other areas of the body in the street, but this process can be accomplished in the better conditions of the hospital, as shown in **Figure 5-32.** To perform fine decon according to OSHA, the staff must have proper training to perform these activities and must be trained to the operations level, although some training to the technician level is preferred. Some

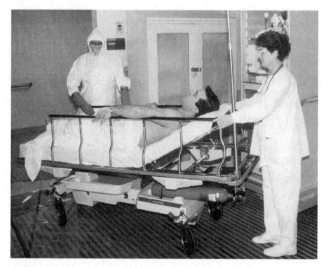

Figure 5-32 Fine decontamination occurs at a hospital and is used to clean the fingernails, ears, and other hard-to-reach areas. *(Courtesy of Baltimore County Fire Department)*

hospitals have designated decon areas with a separate entrance, separate ventilation system, and a holding tank for runoff control. Through the LEPC plan, hospitals are required to be identified to accomplish decontamination of patients. Every hospital should have a policy to cover a walk-in contaminated patient, and emergency responders should test these plans on a regular basis. Regular training and interface is required to keep this type of system functioning well within a community. Responders should ensure this capability is well run and trained, as they are likely candidates to become patients one day.

Mass Decontamination

Most HAZMAT teams' decontamination setup is designed primarily to decontaminate team members. It is designed to handle two to four people dressed in chemical protective clothing. Although most of these systems can handle civilians, they are intended to handle the needs of cleaning responders' PPE. When decontaminating civilians, several issues arise: clothing, privacy, weather impact, and inconvenience. Mass decontamination requires that plans must be developed to handle thousands of people. Most of the mass decontamination issues arise when terrorism events are discussed. More information on possible terrorism scenarios is provided in Chapter 7. If victims are contaminated with military nerve or blister agents, speed is of the essence. If decontamination is not performed on symptomatic contaminated patients, their chance for survival is slim. The hardest part of this process is identifying those who need decontamination and those who need **psychological decontamination.** If victims are symptomatic and a credible threat is presented, decontamination should be accomplished expeditiously. First arriving units need to establish this decontamination process. The best decontamination solution is water, followed by soap.

SAFETY The use of a bleach and water solution in terrorism decontamination is no longer considered an acceptable practice.

Studies from the military have shown that the use of bleach can actually cause more deaths and injuries among contaminated victims. Decontamination is chemical specific and unless responders can identify the specific chemical compound, they should use a simple soap and water solution. A simple hoseline can be used to accomplish emergency victim decontamination; as more assistance arrives, the system can become more elaborate. When setting up the decontamination process it is important that patient backup be avoided. If there are delays, make sure that patients wait in water. After the patients exit the decontamination area,

they should be conducted to holding areas. It is important to keep symptomatic patients away from nonsymptomatic patients. If these patients are mixed, then the nonsymptomatic victims will tend to pick up the symptoms due to psychological reasons. Examples of mass decontamination setup are provided in **Figure 5-33.**

Decontamination Process

Decontamination procedures may vary slightly, but most organizations have developed similar procedures. In **Figure 5-34** the photos show the basic process and outline the following steps: tool drop, gross decon, formal decon, PPE removal, SCBA removal, clothing removal, body wash (emergency only), clothing removal, dry off, and medical evaluation. The variations include the use of shower systems, tents, number and location of steps, and personnel. The method of runoff collection also varies from team to team. **Figure 5-34** depicts the use of a 3-inch hoseline that is cut and coupled to provide a rectangular setup; it includes a shower wand manifold. For most situations that involve solids and liquids a minimum type of decon should be set up. The HAZWOPER regulation requires that the IC have a plan for decontamination, which can vary from no setup through a multistep process as depicted in the decon photo series. For a material such as a technical grade pesticide, a full setup would be recommended with all stations established. For gases, in reality, no decon is required, but a minimum setup should be established prior to entry in the event of an emergency and for psychological purposes. No matter the type of size of the setup that will be used, it should be organized and ready prior to any entry into a hazard area.

METHODS OF DECONTAMINATION

Table 5-6 lists general categories of decontamination that apply to humans, equipment, and the environment, although they may not apply directly to all three. Be sure to consult with the HAZMAT team or a chemist prior to using any of these methods on a human, as they may be more dangerous than the contaminant.

ABSORPTION

The spilled material is picked up by the absorbent material, much like a sponge. Common absorbent materials include ground-up newspaper, clay, kitty

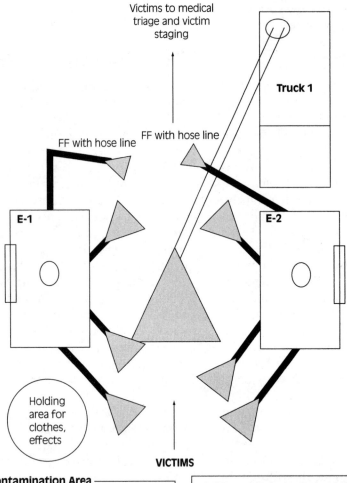

BASIC PLAN
Baltimore County Fire Department
Hazardous Materials Response Team
Mass Casualty Decontamination

Victims to medical triage and victim staging

Truck 1

FF with hose line

FF with hose line

FF with hose line

E-1

E-2

Holding area for clothes, effects

VICTIMS

Contamination Area

Victims are decontaminated with nozzles on the rear, and side discharges. Deck guns and a ladder pipe are also used. At the end of the line 2 FF's using 1 3/4" hose lines complete a gross decon of the victims. All victims who are contaminated and require decon should wait in the water until hosed off. Once done, ALL victims should be held for further examination or questioning.

Figure 5-33   Mass decon setups.

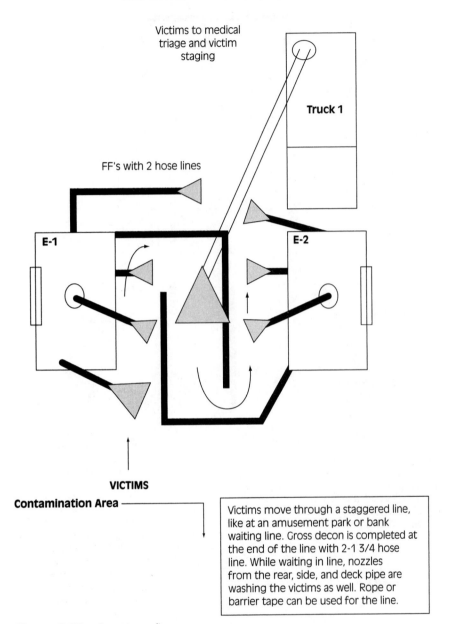

ADVANCED PLAN
Baltimore County Fire Department
Hazardous Materials Response Team
Mass Casualty Decontamination

Victims to medical triage and victim staging

Truck 1

FF's with 2 hose lines

E-1

E-2

VICTIMS

Contamination Area

Victims move through a staggered line, like at an amusement park or bank waiting line. Gross decon is completed at the end of the line with 2-1 3/4 hose line. While waiting in line, nozzles from the rear, side, and deck pipe are washing the victims as well. Rope or barrier tape can be used for the line.

Figure 5-33 *(continued)*

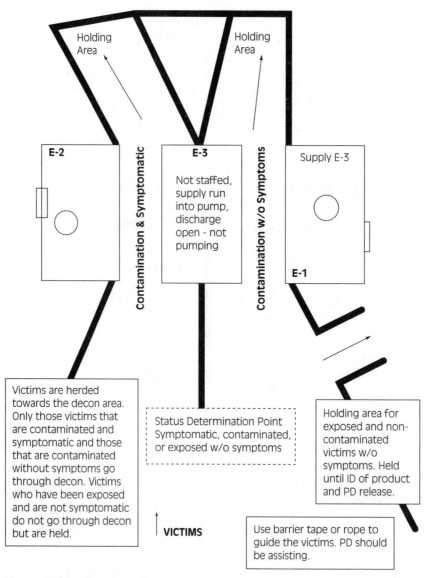

Herding Guide
Baltimore County Fire Department
Hazardous Materials Response Team

Holding Area

Holding Area

E-2

Contamination & Symptomatic

E-3

Not staffed, supply run into pump, discharge open - not pumping

Contamination w/o Symptoms

Supply E-3

E-1

Victims are herded towards the decon area. Only those victims that are contaminated and symptomatic and those that are contaminated without symptoms go through decon. Victims who have been exposed and are not symptomatic do not go through decon but are held.

Status Determination Point Symptomatic, contaminated, or exposed w/o symptoms

Holding area for exposed and non-contaminated victims w/o symptoms. Held until ID of product and PD release.

VICTIMS

Use barrier tape or rope to guide the victims. PD should be assisting.

Figure 5-33 *(continued)*

Water Flow for Herding Guide
Baltimore County Fire Department
Hazardous Materials Response Team

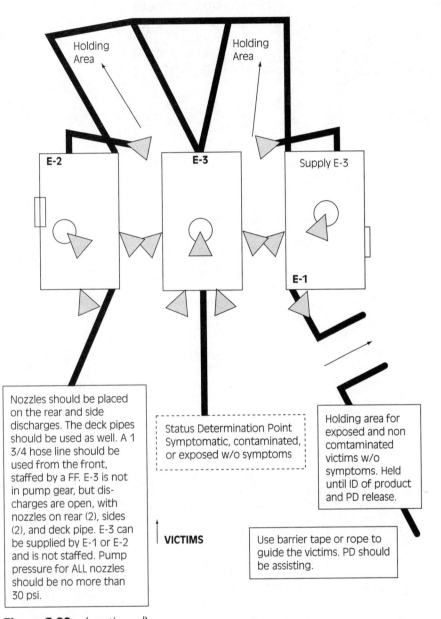

Figure 5-33 *(continued)*

litter, sawdust, charcoal, and poly fiber. The absorbed material does not change, and the volume of material may increase. The whole mixture should be treated as waste and will not change any characteristics such as flammability. Compatibility needs to be researched prior to using these materials, as some materials could cause a reaction.

ADSORPTION

With this method the material bonds to the outside of the adsorption medium. Activated carbon and sand are the most common adsorbents and should

be available locally at a chemical facility. Many chemical facilities have drums, bags, or totes of activated carbon set up to receive liquids within a process, and those materials may be available in the event of an emergency.

COVERING

This option is useful for solid materials, but could also be used for liquids as well. As long as the chemical is compatible with the covering material and will not permeate the material, it would be useful to cover a spill. Once the spill area is covered, a

JOB PERFORMANCE TASK

Figure 5-34 Steps in the Decontamination Process

A The tool drop is usually the first stage in the decon setup. Tools are placed for the use of another entry team or may be collected for later decontamination.

B In most systems gross decon is the next step. Some response teams use showers or hoselines to accomplish this task.

C The next step is formal decon, which may involve two stations in which the responder is rinsed, scrubbed, and rinsed again. The solution depends on the chemical. This is the first step in which other responders may assist.

D After the wet portion the entry team removes PPE with the assistance of other responders.

E After the removal of the PPE, the entry team remove SCBA. It is usually placed into some type of containment system for later cleaning.

(continued)

JOB PERFORMANCE TASK

Figure 5-32 (continued)

F In some systems a fourth washing area is established for body washing. Some teams may set up a tent for this purpose.

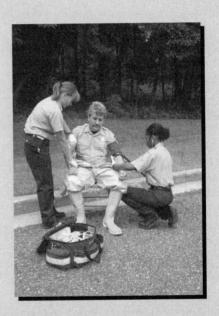

G The last step is medical evaluation and rehydration.

TABLE 5-6

	Methods of Decontamination		
Type	Used on humans	Used on equipment	Used on environment
Absorption	Yes	Yes	Yes
Adsorption	Yes	Yes	Yes
Covering	No	Yes	Yes
Dilution	Yes	Yes	Yes
Disinfection	Yes	Yes	Yes
Disposal	No	Yes	Yes
Emulsification	Yes	Yes	Yes
Neutralization	Under specific conditions	Yes	Yes
Overpacking	No	Yes	Yes
Removal	Yes	Yes	Yes
Solidification	No	Yes	Yes
Vacuuming	Yes	Yes	Yes
Vapor Dispersion	Yes	Yes	Yes

method of securing the seal must be used to prevent the cover from blowing away.

DILUTION

The ability to dilute a contaminant depends on the chemical structure of the spilled material. Dilution is not effective with all chemicals. Imagine being coated with used motor oil: Flushing with water will not remove all of the contamination. When dealing with corrosives, large quantities of water are required, sometimes hundreds of thousands of gallons. Some recent military studies have determined that plain water can be effective and that some bleach and water solutions can cause greater problems for patients. When time and resources permit, soap can be added as this option as it is shown to be of additional benefit in the decon process.

DISINFECTION

When dealing with humans, a 0.5 percent bleach and water solution can be used for some etiological contaminants. Initially, bleach solutions were believed to be effective for biological contaminants as well, but researchers have determined that plain water is just as effective for humans. For equipment or the environment, disinfection can be accomplished using a higher percentage of bleach or other chemical.

DISPOSAL

In limited cases it may not be possible to decontaminate a tool cost effectively. In those cases disposal is the only option. Contaminated soil in many cases is dug up and disposed of either at an incinerator or hazardous waste landfill.

EMULSIFICATION

This is a form of neutralization, or chemically altering a contaminant. Many emulsifiers are sold for fuel and oil products to chemically break down the fuels into other components. The individual components may still be harmful to the environment, and the use of emulsifiers may be illegal in some states, except under extreme conditions. Emulsifiers are useful when the contaminant is in an unusual area, such as in a sump pump in a residential basement, and would be difficult to remove otherwise. Prior to using any emulsification agent, check with your local or state environmental authority as many of these agents are prohibited. In some cases permission for extraordinary use may be granted.

NEUTRALIZATION

This method is usually reserved for corrosive materials but also is used here to describe the procedure to reduce the toxicity of a poisonous material. It is highly recommended that responders consult with a chemist prior to performing this activity. Many chemicals used to reduce this risk are in themselves hazardous, though usually in a much lesser fashion than the contaminant. An emergency responder may avoid mixing a neutralization solution using sodium hydroxide as it is very corrosive, but with some chemicals it is extremely useful in small amounts in reducing the hazard of some of the contaminant. In general, neutralization is exothermic and can produce violent reactions, although these are extremely rare.

OVERPACKING

This objective will be dependent on the material involved and the size of the container. Overpacking entails placing the material into another container, usually a larger drum. Contaminated clothes may be placed into a 55-gallon drum known as an *overpack drum*.

REMOVAL

This option is reserved for soil in most cases, although equipment or tools could also removed from the area to eliminate the contamination. In most cases the contaminant is removed and overpacked.

SOLIDIFICATION

A solidification agent is used to make a liquid material into a solid and may alter the spilled agent. In some cases, however, the solidification agent will not have any effect on the agent and will reduce the hazard and enable later sampling. Chemical compatibility needs to be confirmed prior to the use of a solidification material.

VACUUMING

For dusts, fibers, or other solid materials, a vacuum would certainly reduce the hazard. An ordinary shop vac would not be recommended for this task, however, as it is not filtered well enough to keep the exhaust from blowing agent back into the air. Use a vacuum that is equipped with a high efficiency particulate accumulator (HEPA) filter. Specialty vacuums are made exclusively to pick up mercury without the mercury vapors being released into the air.

VAPOR DISPERSION

General ventilation can be used to decontaminate or clear a building. When humans are involved, fresh air can be a decontaminating solution. The use of Positive Pressure Ventilation (PPV) fans can assist in the removal of contaminants in a building, but levels of carbon monoxide must be monitored. It is possible to add a vapor dispersant into a building to speed up the breakdown of the contaminant, but consultation with a chemist is recommended.

SUMMARY

Protective actions are used for a variety of purposes to protect the public and, in many cases, the responders. The management of a chemical release is not an easy task, as multiple agencies are often involved in accomplishing a variety of tasks at the same time. Unlike most fires or accidents, political pressure may be associated with a chemical release, especially if there is the potential that thousands of people may require evacuation and relocation. The determination for decontamination can also be a difficult decision, as most departments do not have the ability to decontaminate large numbers of victims. Choosing the method of decontamination is often complex, as there are many possibilities. The best options for protection are to provide prompt isolation and prevent further escalation of the exposed population. For those persons who are contaminated, a water-based decontamination is usually recommended.

KEY TERMS

8-step process[©] Management system used to organize the response to a chemical incident. The elements are: detect, estimate the harm, choose an objective, identify the action, do the best you can, and evaluate your progress.

Bung Cap of a drum. The cap varies in size but the most common is ¾ to 2½ inches.

Containment system Provided as a backup system in the event of a tank failure. If the tank fails, the containment system is designed to hold the released product.

DECIDE Management system used to organize the response to a chemical incident. The factors of DECIDE are: detect, estimate, choose, identify, do the best, and evaluate.

Decontamination Physical removal of contaminants (chemicals) from people, equipment, or the environment; most often used to describe the process of cleaning to remove chemicals from a person.

Emergency decon Rapid removal of material from a person who requires immediate cleaning. Most emergency decon setups use a single hoseline to perform a quick, gross decon.

Emergency response planning (ERP) Levels that are used for planning purposes and are usually associated with preplanning for evacuation zones.

Evacuation Movement of people from an area, usually their homes, to another area that is considered to be safe. Persons are evacuated when they are no longer safe in their current area.

Fine decon More detailed form of decontamination, usually performed at a hospital by staff who are trained and equipped to perform decon procedures.

Formal decon Washing and scrubbing portion of the decontamination process. The process is usually repeated and is performed by a decon crew.

GEDAPER© Management system used to organize the response to a chemical incident. The factors are: gather information, estimate potential, determine goals, assess tactical options, plan, evaluate, and review.

Gross decon Step in the decontamination process that removes the majority of the chemicals through a flushing of the person. The gross washing employs large amounts of water and is usually done by the individual or a partner.

Hazard area Term used to describe the dangerous area around a release; also called the isolation area or hot zone.

Hot, warm, and cold zones All zones of isolation used at chemical releases from the most dangerous to the safest areas.

Isolation area Area in which citizens and responders are not allowed; may later become the hot zone/sector as the incident evolves. This is the minimum area that should be established at any chemical spill.

Overpacked Response action that involves placing a leaking drum (or container) into another drum. Some drums are made specifically to be used as overpack drums and are oversized to handle a normal-sized drum.

Psychological decontamination Used to make patients who are not contaminated feel better, producing psychological cleaning.

Responsible party Person or company responsible for the spill or release, who is ultimately responsible for the cleanup of the incident.

Sector See *zone*.

Shearing effect Tearing of the molecules, which can be equated to being ripped apart.

Shelter in place Form of isolation that provides a level of protection while leaving people in place, usually in their homes. People are usually sheltered in place when they may be placed in further danger by evacuation.

Sick building Term associated with indoor air quality; occupants in a sick building become ill as a result of chemicals in and around the structure.

Sick building chemical Substances that cause health problems for occupants of a structure.

Span of control Number of workers that one commander can effectively manage. When the number of workers exceeds the span of control, the supervision must be divided among more commanders.

Zone Area established and identified for a specific reason; typically a hazard exists within the zone. Zones are usually referred to as a hot, warm, and cold to provide an indication of the expected hazard in each zone. Some agencies use the term *sector*.

REVIEW QUESTIONS

1. What is the first priority of any incident management system?

2. What zone or sector should be established first?

3. When dealing with a flammable liquid spill, what criteria could be used to establish the hot zone?

4. What level incident would be used to describe a chemical release in which 2,000 people were evacuated?

5. What is the minimum decontamination that should be set up for every incident?

6. What are the four types of decontamination?

7. In what step of decon are hospitals involved?

8. To decide to make a rescue depends on what information?

9. What causes the most common breach in a 55-gallon drum?

10. If fire has reached the cargo portion of a truck carrying DOT Class 1.1 materials, what tactics are used to fight the fire?

11. How can air monitors be used in the determination of hazard zones?

ENDNOTES

[1] In most states the fire department IC is in charge of the incident until the IC relinquishes command to another agency. Check your local and state regulations for more information. Although some responders leave when the cleanup contractor begins work, a safer policy may be to maintain a presence as long as the scene is considered hazardous. The emergency responders are solely responsible for public safety, a responsibility that cannot be delegated. The fact that a group has been "hired" to work in a hazardous environment does not relieve the responders of its responsibility. In

many cases the contractor will use appropriate measures to protect employees and the adjacent community; in those cases the responders may choose not to remain on site. In cases where the contractor does not have the ability to protect workers through the means of air monitoring, as an example, someone must provide oversight. In some states this can be handled by the state or local environmental agency. But ultimately the responsibility for public safety is a emergency response function, and a contractor is part of that "public" we are sworn to protect.

[2]Any location such as a storm drain, ditch, or culvert that eventually leads to a waterway is defined as a waterway.

Suggested Readings

Lesak, David, *Hazardous Materials Strategies and Tactics,* Prentice Hall, New Jersey, 1998.

Noll, Gregory, Michael Hildebrand, and James Yvorra, *Hazardous Materials: Managing the Incident,* Fire Protection Publications, Oklahoma University, 1995.

Schnepp, Rob, and Paul Gantt, *Hazardous Materials: Regulations, Response, and Site Operations,* Delmar, a division of Thomson Learning, Albany, NY, 1999.

Smeby, L. Charles (editor), *Hazardous Materials Response Handbook,* 3rd ed., National Fire Protection Association, 1997.

CHAPTER 6

PRODUCT CONTROL AND AIR MONITORING

STREET STORY

In January 1983 I was faced with the realization of how little I knew about chemical emergencies. It was a hard lesson to learn. I was called at home around 4 A.M. to respond to the downtown area of my city for a fire that involved chemicals. It was reported that several firefighters had been injured. En route I was in radio contact with several officers on the scene and was informed that there had been an explosion and that four firefighters had been burned.

Once on the scene, I learned that a firefighter who had driven a car in my wedding party was being sent to a burn center in New York City. Not only was I dealing with a difficult incident, but it also now had an emotional aspect tied to it. Even with my limited knowledge, I was given the job of informing and advising the incident commander of the potential hazards involved with several of the chemicals that had been identified. The fire was caused by an overpressurization of a reactor vessel. Because of the explosion, no product control was initiated, and to compound the problem, the property was adjacent to the Long Island Sound. This incident was causing environmental damage to the land, water, and air, and four firefighters had already been injured. My cadre of tools included two reference books and sixteen hours of hazardous materials training. After several difficult hours of trying to identify, evaluate, and provide useful information to both the IC and the hospital, I came to the conscious decision that I simply did not have the right tools nor the required knowledge to be very helpful.

This one incident changed the way I would respond to chemical emergencies for the rest of my life. This incident taught me the importance of good identification skills, the requirement to understand every term as it relates to the hazards of the product, and the absolute need for air monitoring devices. To keep our troops safe, being armed with knowledge is the only way to face any type of emergency.

—Street Story by Frank Docimo, Special Operations Officer, Turn of River Fire Department, Turn of River, Connecticut

OBJECTIVES

After completing this chapter the reader should be able to identify or explain:

- Equipment that can determine hazardous environments and isolation areas (O)
- Available defensive operations for a release (O)
- The various methods of damming, diking, diverting, and other defensive operations (O)
- Detection equipment available for the first responder and equipment that a HAZMAT team may carry that could be used at a chemical release. (O+)

INTRODUCTION

The use of product control techniques can provide a quick reduction in damage done to the community and the environment. The reduction of surface area over which the product has spread provides a direct reduction in the danger to responders and the community. By implementing these types of measures, first responders can provide an extreme advantage to the overall mitigation of the incident. The use of air monitoring devices is becoming more commonplace, and many first responders have some type of air monitoring device available to them. This section provides a brief overview of this complex subject and concentrates on the knowledge first responders should have. To ensure responder safety, basic air monitoring must be accomplished, even for natural gas leaks, gasoline spills, or any other incident in which responders may be exposed to hazardous materials.

DEFENSIVE OPERATIONS

The NFPA identifies two large categories that fit into defensive operations at the first responder level: containment and confinement. Both of these can be combined into product control, which is one task that persons trained to the operations level can perform as long as they perform these duties away from the hazard area. To perform defensive operations, responders do not need to get close to the actual spill. The activities that fit into defensive operations are: **absorption, diking, damming, diverting, retention, dilution, vapor dispersion, vapor suppression,** and the use of **remote shutoffs.** The type and level of PPE required for each of these tasks varies with the chemical involved, but a minimum would be TOG and SCBA. If making a dam a

considerable distance away from the chemical release and if no contact with the material is possible, a lesser level could be used. The tasks described in this chapter are among common tasks assigned to responders trained to the operations level. These critical tasks can protect the community and the environment. It is not necessary for first responders to carry all of the equipment cited in this chapter, but they must know its location and how to improvise, adapt, and overcome.

Absorption

First responders are often asked to handle absorption of a spilled material, and on a daily basis they probably absorb a large quantity of fuel and oil products using a product similar to the one depicted in **Figure 6-2.** Knowing the type of absorbent materials that a department carries is important. Some absorbent materials will not pick up water, while others absorb any liquid with which they come in contact. When trying to remove fuels from water, using an absorbent that does not pick up water is essential. Absorbents vary from clay kitty litter to ground-up newspaper, corn husks, and synthetic fibers. Their weight and absorbent capabilities vary, and not all can be used interchangeably. They may

Figure 6-2 A common defensive action is the application of absorbent to a spilled material.

Figure 6-3 Absorbent pads will pick up oil from the water but will not absorb the water, allowing them to float.

Figure 6-4 A screen fence can be used to collect the absorbent material floating on the water, and the absorbent will collect the spilled material. The water flows through the fence below the absorbent.

Figure 6-5 When time permits, more advanced methods of damming a waterway can be used. This system is more advanced than most but takes time and resources to accomplish. *(Courtesy of the Maryland Department of the Environment)*

Figure 6-6 The use of natural surroundings and materials assists in the construction of a filter fence used to skim material from the top of the water. In this case a snow fence is used to filter the water and hold the absorbent material. A fallen tree adds support to the operation. In some larger spills HAZMAT crews have felled trees to assist in this process. *(Courtesy of the Maryland Department of the Environment)*

be available in loose bulk form or as pads, such as those shown in **Figure 6-3,** or as large rolls or in boom style. The recommended method to hold the absorbent material is to construct a filter fence such as the one shown in **Figure 6-4.** The filter fence can be made of "rat wire" or other small-weave fencing material. Wooden stakes can be used to hold the fence in place; in rocky soil metal stakes are recommended. Some examples of how to use common materials are shown in **Figure 6-5** and **Figure 6-6.**

Always construct multiple systems so that in the event of a failure the spilled material can be caught by a backup system. Also available on the market are solidification agents that encapsulate the spill and microbiological oil-eating bugs. Both of these agents are expensive, and absorption materials are not generally used by emergency services due to their cost and the quantity required. It is best to carry a variety of absorbent products in several different styles.

Diking and Damming

The quantity spilled will determine to what extent these operations are performed. The actions required at a 50-gallon spill differ from those of a 500,000-gallon release. The two possible locations for diking, damming, and diverting operations are a spill on the ground or a spill on a waterway. A spill on the ground is much easier to control than one on the water, but the principles are the same. When creating a dike or a dam, you are either stopping the flow of the material or keeping it from a specific area. Except for small streams it is nearly impossible to stop a body of water from flowing. If the spilled material is soluble with water and the waterway has a good flow, it is impossible to collect the spilled material unless responders have heavy equipment and a large area available to create a very large dam.

Commonly first responders use earth, sand, or rocks for dikes or dams, depending on what is available. Having local contacts available around the clock for these materials is important, and some jurisdictions have emergency contacts for dump truck loads of sand or dirt. Local or state highway departments are a good place to start when these items are required. Heavy equipment such as front-end loaders or track hoes can also be used to create dams or dikes. When constructing a dam or a dike,

it is best to follow the rule of three and construct three setups. Start at the furthest point away from the spill and work back toward the spill. It may be necessary to establish the first containment miles away from the spill. If you start near the spill, you are likely to be overtaken by the spill and be forced to play catch-up in additional PPE.

Two basic types of dams can be created: overflow and underflow dams, as shown in **Figure 6-7.** The specific gravity of the spilled material, or whether the material is floating on top of the water or traveling on the bottom, will determine the type of dam that must be constructed. An overflow dam allows the water to overflow the dam and contain the spilled material at the base of the dam. An underflow dam allows the water to flow under the dam and collect the material on top of the water. Most spilled materials will be floating on the top of the water and require the use of an underflow dam. The materials required for either style dam are shovels, dirt or sand, and pipes. Pipes should

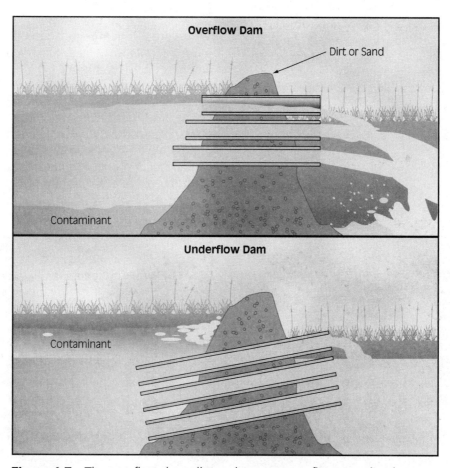

Figure 6-7 The overflow dam allows clean water to flow over the dam, and collects the spilled material at the base of the dam. The underflow dam collects the spilled material on top and allows the water to flow through the bottom of the dam.

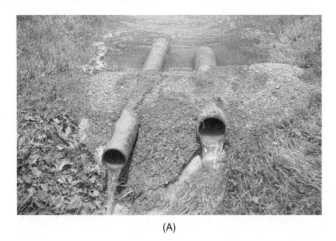

(A)

(B)

Figure 6-8 (A) Overflow dam. (B) Underflow dam.

be 4 inches in diameter at minimum, and larger is better. The higher the flow or the more water, the bigger the pipe and the more pipes that are required. A good rule of thumb is to have enough pipes to cover two-thirds of the width of the waterway, which would be the minimum amount required. Standard PVC pipe is adequate and for emergency situations, the hard sleeves (suction) on a fire engine company are a good substitute. The photos in **Figure 6-8** and **Figure 6-9** show the process of setting up both styles of dams.

For an overflow dam, one tendency is to establish a dam, place the pipes on top, and consider the dam fin-

Figure 6-9 First responders have equipment such as hard sleeves to establish over- and underflow dams.

ished. It is best to set the pipes on top of the dam and then continue with a layer of dirt/sand on top of the pipes to avoid the possibility that the water will erode the dam and the containment will collapse without this extra layer of dirt/sand. **Figure 6-10** is another example of how to construct an underflow dam.

Diverting

To divert a running spill, almost anything can be used, such as dirt, sand, absorbent material, tarps, and hoselines (see **Figure 6-11**). The most common use of diversion is to keep a running spill from entering a storm drain. A ring of dirt around the storm drain will keep the spilled material out but does not stop the flow of the product. It may be impossible to stop the flow of a large quantity of the spilled material, but it would be possible to keep it from the storm drain, as the protective measures in **Figure 6-12** demonstrate. Diversion can even be

Figure 6-10 Another method used in the construction of an underflow dam. Note the number of pipes that have been used to move the water. *(Courtesy of the Maryland Department of the Environment)*

Figure 6-11 Plastic filled with sand is used to create a berm to hold back a potential spill from a catastrophic release from the tank truck. The larger the potential, the more substantial the device should be. This tank truck could hold between 9,000 and 10,000 gallons, which if dumped at one time would probably overtake this system. *(Courtesy of the Maryland Department of the Environment)*

Figure 6-12 If it cannot be controlled, a spill should be diverted around areas such as sewers.

Figure 6-13 This harbor boom has a 12- to 18-inch weighted skirt hanging below the waterline to hold back floating product. It is called a harbor boom as it is commonly used to boom around a ship while it is docked in the harbor. *(Courtesy of the Maryland Department of the Environment)*

Figure 6-14 Another use of harbor boom and a floating skimmer shown in the upper left of the boomed area. The skimmer pulls floating liquids, such as fuel oil, off the water for recovery. This type of equipment is usually provided by a cleanup contractor. *(Courtesy of the Maryland Department of the Environment)*

Figure 6-15 This floating boom collects the spilled material that is floating on top of the water. Like pads, the boom does not absorb water and should stay afloat even when saturated with the spilled material.

used for spills on the water. By using a solid boom, such as a harbor boom, material can be diverted into another area or even contained. This type of boom is shown in **Figure 6-13** and **Figure 6-14.**

Another method of diversion is to dig a trench alongside the waterway to collect the spilled material off the top of the water. For large spills this is effective if there is a collection vehicle such as vacuum truck standing by to collect the spilled material as it spills into the diversion. These types of vehicles are available from hazardous waste cleanup contractors. The most common method is to use a floating absorbent boom, which will absorb the material off the water, as shown in **Figure 6-15.**

Retention

The most common method of retention is digging a hole, either by hand or by machine like the track hoe shown in **Figure 6-16.** To catch a running spill, the best method is to dig a hole large enough to collect the spill and then divert the spill into that containment area.

NOTE The ability to create a large enough containment area is paramount to success.

If unable to build a large enough containment, responders should concentrate on building several in a row to collect the material. As with dikes and dams it is best to start the furthest from the spill and create several other areas closer to the spill.

SAFETY When digging retention areas, be aware of underground utilities such as water, electric, sewer, cable TV, phone, or natural gas.

Dilution

Much like dilution in decontamination, this is not always the solution to pollution. If a water-soluble material is in a waterway, then for all purposes it is being diluted. If the waterway is small, there may be occasions when the fire department may be needed to add some water to the spilled materi-

als. Simple flushing of a material into a waterway is no longer an acceptable tactic, and it should not be considered. Items such as fuels, oils, or other hydrocarbons are not usually water soluble and cannot be diluted by water. Dilution may be an effective response for some specific chemicals when combined with other containment tactics.

Vapor Dispersion

This topic can create some confusion for responders. By their nature, firefighters like to use water, usually in large quantities. In many occasions the application of water will make the situation safer than without a water spray. In incidents involving a severe threat to life, such as an adjacent nursing home that cannot be evacuated, using a water spray to disperse vapors would be a good idea. If, on the other hand, the release is occurring in a remote area away from any population, a water spray to disperse vapors, such as the situation shown in **Figure 6-17,** is not necessary. To be effective the material being released must be water soluble or the vapor cloud must be able to be moved by the water streams. Although the DOT ERG suggests for many spills to "use water spray to reduce vapors or divert vapor cloud," this course is not recommended except in life-threatening scenarios. Some departments use water sprays at natural gas leaks, which is not necessary as it further complicates the repair of the leak. Many leaks occur in a hole that was created, and the use of water spray fills the hole with water and makes repair dangerous and complicated. In addition, the use of water may actually knock down vapors that normally rise up and dissipate quickly.

Figure 6-16 Small retention areas can be dug by hand, but larger spills require larger equipment. Responders should have established contacts for such equipment.

Figure 6-17 The use of a water spray is not necessary unless the vapor cloud is threatening the public. In some cases it creates more hazards than those it may eliminate.

When using a water spray, you may knock down the vapors, but you also may create another substance. When anhydrous ammonia is released into the air and a water spray is applied, the resulting runoff is ammonium hydroxide, an extremely corrosive liquid that itself would require containment and cleanup. Both natural gas and propane are not water soluble, and the use of water vapor simply relocate the vapors to another area.

> **NOTE** It is important to know what material is leaking prior to the application of water, as the water spray may cause more problems. It is very important to consult with the HAZMAT Team prior to utilizing this tactic.

Vapor Suppression

With the use of firefighting foams, vapor suppression is another tactic that first responders can use. The type of material spilled will depend on the type of foam that should be used, as not all foams will work on all products. See Chapter 5 and *The Firefighter's Handbook* for more information regarding foam and its application. The use of foam to suppress vapors is generally limited to flammable liquid spills, but some chemical foams are available for other types of materials. Before applying foam, responders should ensure that the material is contained, that the application of foam will not cause any further problems, and that the foam is compatible with the spilled material. As with foam to extinguish a fire, responders should make sure they have enough foam stockpiled to keep a layer of foam on the spilled material.

Remote Shutoffs

Shutting off valves is not usually a tactic used by first responders, but there are some exceptions to this rule. On most tank trucks and at some fixed facilities, remote shutoffs can be operated by the first responders. Each type of truck is different, but in general an emergency shutoff is located behind the driver's side of the cab, and another is near the control valves. The shutoffs are mechanically operated and may be self-operating in the event of a fire. They are usually well marked as "emergency shutoff" and in an easy-to-find location. At some facilities, typically those with loading racks, there may be a remote shutoff near the entrance so that in the event of an emergency first responders can shut off the flow of product.

AIR MONITORING AT THE FIRST RESPONDER OPERATIONS LEVEL

The many responders purchasing detectors to assist with carbon monoxide alarms are realizing associated benefits. Depending on the type purchased, it may also be used to detect flammable gases, oxygen levels, and one or two toxic gases. An example is shown in **Figure 6-18.** Many responders use these instruments for flammable gas leaks and for confined space entries. The use of air monitoring is one of the most crucial components in the response to chemical incidents. Air monitoring can keep responders alive, but first responders must have the proper training and equipment to ensure their safety. When responding to situations that involve unknown materials, HAZMAT responders need pH detection for corrosives, a combustible gas detector for the fire risks, and a photoionization detector (PID) for

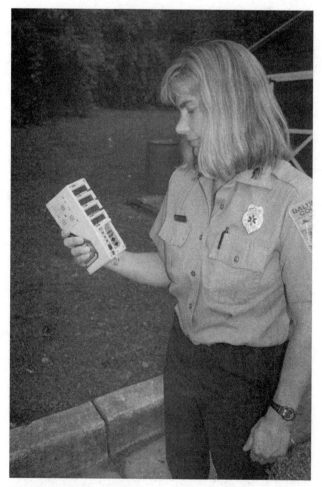

Figure 6-18 Air monitors, especially those called three- and four-gas instruments, are being used more commonly by first responders.

the toxic risks. Unfortunately many responders only rely on one or two of these detectors. The most common detector is a three-, four-, or five-gas instrument, and the other instruments are not available or not used. By not using a method of detecting toxic materials, responders (and the public) can be lulled into a false sense of security. Most combustible gas detectors are not made or designed to measure toxic gases in air, and they only read those types of gases when they become fire hazards. Many flammable gases are very toxic at considerably lower levels than their lower explosive limit. Some new combustible gas detectors measure in both ppm and percent of the LEL, although the ppm range usually starts at 50 ppm or higher. Many first responders have purchased air monitors, but may not have a full understanding of how they work.

> **NOTE** The use of air monitoring and sampling equipment can be complicated and requires practice.

This text will provide information related to air monitoring and its importance in emergency response.

Most responders have a basic understanding of combustible gas indicators, but even that knowledge can be limited. Our society is becoming much more sophisticated and the citizens (our customers) expect more. This section presents some concepts on air monitoring, monitoring strategies, information on how the monitors work, and their uses. The field of air monitoring technology is ever changing and new technology emerges each year, although the basic principles of safe decision making remain the same. When purchasing air monitors it is important to understand the basic features of the instruments so that the purchase benefits the organization.

> **NOTE** The general rule when purchasing an air monitor is to figure one-fourth of the purchase price into the annual budget for upkeep and repairs. It is necessary to keep the instrument calibrated and repaired on a regular basis.

Regulations and Standards

Within OSHA's HAZWOPER Regulation (29 CFR 1910.120) are few specific requirements for air monitoring, even for the HAZMAT technician, but air monitoring is the principle safety element throughout the document. OSHA specifies that the IC must identify and classify the hazards that are

present on a site. The use of air monitoring is the primary key to fulfilling this obligation, as is the staffing of a HAZMAT technician trained to use detection devices. The NFPA 472 document also has some requirements for air monitoring, but like the OSHA regulation, these are fairly generic. There are no requirements at the operations level for the use of air monitors. It is important for first responders to understand how monitors work and what their deficiencies are, as they may place their lives on the line depending on the readings of an air monitor. First responders are more affected by the confined space regulation as it applies to air monitors than they are by the HAZMAT regulation. Unfortunately, it is likely that a lack of understanding and/or maintenance will play a factor in future responder fatalities and injuries based on the large number of departments using these instruments.

Air Monitor Configurations

Most responders purchase an instrument that is known as a three-, four-, or five-gas instrument, commonly referred to as a **multi-gas detector**, such as the one shown in **Figure 6-19.** In other words it samples for three to five gases. A typical detector can monitor oxygen levels, LEL, and the presence of two or three toxic gases. Typical toxics are carbon monoxide (CO) and hydrogen sulfide (H_2S), and if a fifth gas can be sampled, most departments

Figure 6-19 These instruments are four-gas instruments and measure LEL, oxygen, carbon monoxide, and hydrogen sulfide.

choose chlorine or sulfur dioxide. For most responders, the CO and H_2S are more than sufficient. Although these instruments are considered expensive, in the grand scheme of things they are inexpensive compared to other detection devices used by HAZMAT teams.

Many responders may purchase a device with little or no training, which can be detrimental to the successful use of the instrument. In addition to obtaining proper training, departments buying a detector must purchase several additional items to keep the instrument functioning. Typically these items are covered in a special "kit" price and include a battery-powered sample pump, calibration gas, and hardware. The calibration gas usually has an expiration date and will require replacement in six months to a year. The sensors in the detector need regular calibration, and the instrument should be turned on and used on a regular basis, for a period specified by the manufacturer. Rechargeable batteries, depending on the charging system, may need replaced on occasion. The sensors themselves will need replacement, although most manufacturers are providing two-year warranties for the sensors. In general the LEL sensor will last four to five years, the oxygen twelve to eighteen months, and the CO and H_2S detectors eighteen to twenty-four months. Other toxic sensors may last six to eighteen months depending on the type. Most of the perceived problems with instruments are created by a lack of maintenance and adequate training.

NOTE No matter what a salesperson says, no device can conclusively identify an unknown material.

METER TERMINOLOGY

To comprehend fully the use of air monitors, the responder must understand the basic terminology that is generic to all monitors and applies across the board. More detailed information related to specific monitors is provided in later sections.

Bump Test

Two terms that need further explanation are bump test and calibration as they are both related. A **bump test,** also known as a field test, entails exposing a monitor to known gases and allowing the monitor to go into alarm and then removing the gas, as shown in **Figure 6-20.** Exposing the monitor to a known gas allows

responders to verify the monitor's response to the gas. Most manufacturers provide bump gas cylinders, which contain the gases required to check their instrument. When bump testing, follow the manufacturer's recommendations. In most cases the bump test is used to ensure that the alarms function as intended and the instrument is reading. By regularly calibrating and bump testing, users can determine how accurate the monitor will be in the field. Some instruments react very well to bump testing and display the levels as provided on the bump gas cylinder, while others will be off slightly. When responding to confined spaces responders are required by OSHA to bump test the instrument prior to entry into the confined space.

Calibration

Calibration is used to determine if a monitor responds accurately to a known quantity of gas. When new sensors are installed, they will usually read higher than they are intended; calibration electronically changes the sensor to read the intended value. As the sensor gets older, it becomes less sensitive and calibration electronically raises the value that the sensor displays. When a sensor fails, it cannot be electronically brought up to the correct value. Regular calibration gives a picture of the expected life of a sensor, as it will deteriorate over time.

The regularity of calibration is subject to great debate, as some departments calibrate daily, others every six months. The only guideline found in the regulations (for anything that requires air monitoring) is in the confined space regulation, which requires calibration according to the manufacturers' recommendations. Most of the written instruction

Figure 6-20 A bump test assures that the instrument responds to known gases. The detector should reach alarm levels to verify that it is functioning.

guides from the manufacturers require calibration before each use. The definition of calibration at this point is also subject to debate, as responders can verify the monitor's accuracy by exposing it to a known quantity of gas, but not perform a full calibration. Most response teams establish a regular schedule of calibration (weekly/monthly) and then perform bump or field checks during an emergency response. Check with the manufacturer as to what calibration/bump test policy it recommends.

Reaction Time

All monitors have a lag time or, as it is better known, reaction time. The reaction time varies depending on whether responders are sampling with a pump. Monitors without a pump operating are in what is called the *diffusion mode* and will generally have a fifteen to thirty second lag time. Monitors with a pump have a typical reaction time of three to five seconds, but even this can vary from manufacturer to manufacturer. Hand-aspirated pumps usually require ten to fifteen pumps to draw in an appropriate sample. Hand-aspirated pumps are not recommended because readings vary widely depending on how the user operates the hand pump. The goal is to provide a given amount of volume across the sensors.

NOTE When using sampling tubing, add one to two seconds of lag time for each foot of hose.

Be sure to follow the manufacturer's recommended lengths of hose to ensure that the pump operates correctly.

Recovery Time

In addition to reaction time, responders should also be aware of the monitor's recovery time. Recovery time is the amount of time that it takes the monitor to clear itself of the air sample. This time is affected by the chemical and physical properties of the sample, the amount of sampling hose, and the amount absorbed by the monitor. Some monitors take an extended period of time to clear if they are exposed to a large quantity of a gas. In some cases the instrument must be taken out of the environment and shut off, then started again. The reaction time will affect the overall recovery time.

Relative Response

When a gas monitor is purchased, it is set to read a specific type of gas, such as methane. If any other gas is sampled, the instrument will usually read the gas but at different values. The term *relative*

response is used to describe the way the monitor reacts to a gas other than the one it was calibrated for. This term is not commonly used by emergency responders. For this reason the LEL instrument is commonly used in a go/no go situation. It cannot identify the gas that is present, and until the gas is identified, the exact amount of gas in the air cannot be determined. To maintain a high level of safety, each person operating an air monitor must have a basic understanding about relative response, as the monitor may lead you into a dangerous situation. Each detector has what is known as a relative response curve that compensates for different types of gases. The monitor's manufacturer has tested the monitor against other gases and has provided the relative response factor, which responders can use to determine the amount of gas present when sampling. The displayed reading is multiplied by the factor to provide the reading for the gas that is actually present. Examples of relative response factors are shown in **Table 6-1.**

At a xylene spill responders using an ISC TMX-412 calibrated for pentane may take a reading 50 percent of the LEL. The response curve factor for xylene is 1.3, which is multiplied by the LEL reading.

$$\text{Detector reading} \times \text{response curve factor} = \text{actual LEL reading}$$

$$50 \times 1.3 = 65$$

Thus the actual percent of LEL is 65 percent, a situation that is worse than the instrument's reading. On the other hand, if responders use the same instrument for a spill of propane, the response curve is 0.8. Using the same scenario, the actual level is 40 percent of the LEL ($50 \times 0.8 = 40$), a safer situation than reported by the detector.

Oxygen Monitors

Oxygen is one of the most important gases for which to sample. Responders need it to survive and the other instruments need it to function correctly. Normal air contains 20.9 percent oxygen; below 19.5 percent is considered oxygen deficient and a health risk, and above 23.5 percent is considered a fire risk. If an oxygen drop is noted on the monitor, responders should assume that one or possibly more than one contaminant is present, causing the reduced oxygen levels. In oxygen-deficient atmospheres, any combustible gas readings will also be deficient and unreliable. In an oxygen-enriched atmosphere, the combustible gas readings will be increased and inaccurate. Most oxygen-enriched atmospheres may be due to a chemical reaction involving oxidizers and typically signal a dangerous situation.

TABLE 6-1

Response Curve Factors (Calibrated to Pentane)		
Gas being sampled	ISC TMX-412 factors	MSA 261 factors
Hydrogen	0.5	0.6
Methane	0.5	0.6
Acetylene	0.7	0.8
Ethylene	0.7	0.8
Ethane	0.7	0.7
Methanol	0.6	0.7
Propane	0.8	0.9
Ethanol	0.8	N/A
Acetone	0.9	1
Butane	0.9	1
Isopropanol	1	1.1
Pentane	1	1
Benzene	1	1.1
Hexane	1.2	1.3
Toluene	1.1	1.2
Styrene	1.1	N/A
Xylene	1.3	N/A

Always use response factors supplied by the manufacturer, and keep in mind these are laboratory estimates. ISC factors are for Industrial Scientific Corporation LEL sensor 1704A1856-200 calibrated with pentane. The MSA factors are for the Mine Safety Appliances MSA 360/361 calibrated with pentane.

Oxygen Monitor Limitations

Most oxygen sensors only last twelve to eighteen months, since they are always working. An oxygen sensor is an electrochemical sensor that has two electrodes within a gel-like material. When oxygen passes through the sensor, it causes a chemical reaction, creating an electrical charge and causing a readout to be provided on the monitor. As long as the sensor is exposed to O_2, it will cause a reaction within the electrolyte solution sealed within the sensor. This is the most commonly replaced sensor, and departments should plan on annual replacement, although some may last longer. Exposures that can damage oxygen sensors are chemicals with lots of oxygen in their molecular structure, including carbon dioxide and strong oxidizing materials, such as chlorine and ozone. The problem with CO_2 is that it is always present in the air, and the higher the percentage, the faster the sensor will deteriorate. The optimal temperature for operation is between 32° and 120° F. Between 0° and 32° F the sensor slows down (the electrolyte is like a slushy, gelatin material); temperatures below 0° F can permanently damage the sensor. The operation depends on absolute atmospheric pressure and calibration to the atmospheric pressure responders will be sampling. Responders should calibrate the sensor especially at the temperature, pressure, and weather conditions of the area they will be sampling in.

Combustible Gas Indicators

Combustible gas indicators (CGIs), also referred to as combustible gas sensors, have been used by the fire service and industry for many years. All types of sensors work, some better than others in different situations. It is important to understand how each of these sensors work, as some departments may only be able to afford to purchase one type. To detect a wide variety of chemicals effectively, it may be necessary to have more than one type of sensor.

Most of the new CGI are used to measure the LEL of the gas for which they are calibrated. The greatest majority of CGIs are calibrated for methane (natural gas) using pentane gas. When calibrated for methane, the CGI sensor will read up to the LEL; some new units will shut down the sensor when the atmosphere exceeds the LEL. This is an important consideration, as the longer the sensor is exposed to an atmosphere above the LEL, the faster it will deteriorate. The CGI reads up to 100 percent of the LEL; thus, if a CGI calibrated for methane reads 100 percent the actual concentration in that area is 5 percent for methane (the LEL for methane is 5 percent). If the CGI reads 50 percent, then the concentration for methane is 2.5 percent. Any flammable gas sample that passes over the sensor will cause a reaction. How much of a reaction depends on the gas. The EPA has established action levels that provide a safe layer expressly because of the relative response curve problem. It has factored in the response curve to provide guidelines, outlined in **Table 6-2.**

Combustible Gas Sensor Types

The basic principle behind most CGIs is that a stream of sampled air passes through the sensor housing, causing a heat increase and conversely creating an electric charge and causing a reading on the instrument. Four types of combustible gas sensors are shown in **Figure 6-21.** It is important that when purchasing or using a monitor that responders know the sensor type. Readings can and do vary between the four, and responders' safety depends on that instrument so they must understand how it works.

WHEATSTONE BRIDGE SENSOR

Formerly a commonly used detector, the **Wheatstone bridge sensor** is essentially a coiled platinum wire in a sealed, heated sensor. When the sample gas passes over the "bridge," it heats up that side of the bridge with a heated platinum filament. If the air sample contains any concentration of a flammable, the platinum filament will get hotter and increase its resistance to an electrical current in a nearly direct proportion to the concentration of flammable content. The change in resistance is compared to a known constant resistance, and the difference is converted to a meter reading. This direct proportion of the concentration is known as a *linear response.* This response is reproducible and can be duplicated. Wheatstone bridge is considered to be older technology, but it is still useful and has a proven track record. Newer and much better Wheatstone bridge CGIs are actually two separate wire coils in the middle of the sensor; both are heated with the sample gas passing over one of the bridges. The sample gas, if flammable, will cause the bridge to heat up, resulting in a difference in the electric resistance between the two bridges. This difference is reported on the readout of the CGI. The advantage of this approach allows for correction for ambient temperature and humidity and for other factors that cause instabilities in the meter reading.

CATALYTIC BEAD

The **catalytic bead** sensor is the most common sensor today. It is similar to the older Wheatstone bridge technology with some new twists. Instead of twists of wire forming a bridge, the catalytic bead sensor uses a bowl-shaped piece of metal that sits in the middle of a straight wire. It is typically coated with a catalytic material that helps burn the gas sample off efficiently. The sensor has two bowls, placed in the same fashion as the Wheatstone bridge, one for sampling and the other for reading the change in the sampling bowl.

METAL OXIDE SENSOR

The **metal oxide sensor (MOS)** is new to the CGI market and has attracted a lot of attention. It is a very sensitive detector, which causes several perceived problems. If used and interpreted correctly, this sensor can provide many clues to many chemical responses. The MOS is a semiconductor in a sealed

TABLE 6-2

EPA Air Monitoring Guidelines		
Atmosphere	**Level**	**Action**
Combustible gas	<10% LEL	Continue to monitor with caution.
	>10%	Evacuate space (OSHA confined space).
	10%–25% LEL	Continue to monitor, but use extreme caution especially as higher levels are found (EPA).
	>25%	Explosion hazard, withdraw immediately
Oxygen	<19.5%	Monitor with SCBA; CGI values are not valid (EPA); OSHA requires SCBA.
	19.5%–25% (OSHA >23.5%)	Continue monitoring with caution; SCBA not needed based on O_2 content only. OSHA requires evacuation of the confined space.
	>25% (EPA)	Explosion hazard, withdraw immediately.

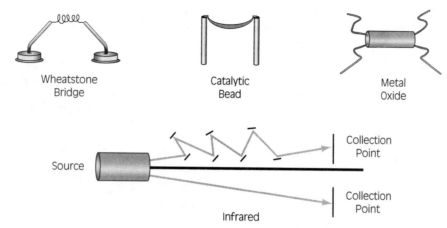

Figure 6-21 From left to right are examples of combustible gas sensors: Wheatstone bridge, catalytic bead, metal oxide, and infrared.

unit with a Wheatstone bridge surrounded by a metal oxide coating. Heater coils provide a constant temperature. When the sample gas passes over the heated bridge, it combines with a pocket of oxygen created from the metal oxide. This reaction causes an electrical change, which produces the reading.

This sensor is very sensitive and will pick up almost anything that crosses it, which can be confusing to some users. The MOS requires regular maintenance and calibration, more so than other types of sensors. Most MOS CGIs do not provide a readout of the percentage but an audible warning or a number from within a range. The range extends from 0 to 50,000 units on a relative scale. If you spill a tablespoon of baby oil on a table and pass a MOS over it, the sensor will report a small reading in the range of 5 to 30. If you take a MOS into a room with 5 percent methane, the reading will be within a range for a flammable gas, probably near 45,000 to 50,000 units. Some monitors allow the MOS to read in percentages of the LEL in addition to the general sensing range of 0 to 50,000 units. The MOS reacts to tiny amounts, which is an outstanding feature of the monitor, as most detectors are not nearly this sensitive. The MOS is typically not the first choice of departments that can afford only one CGI, because it can produce somewhat erratic readings. However, the MOS sensor does have great applicability for a HAZMAT team or a department that already has another type of LEL sensor.

INFRARED SENSORS

The **infrared sensor** is new to emergency response. It uses a hot wire to produce a broad range of light wavelengths and a filter to obtain the desired wavelength; a detection device on the other side of the sensor housing measures that light. The light that is emitted from the hot wire is split, one through the

filter, the other to the detection device to be used a reference source. When a gas is sent into the sample chamber, the gas molecules will absorb some of the infrared light and this light will not reach the detection device, creating a difference in the amount of light reaching the two detectors. The amount of light reaching the detection device is compared and a comparable reading is provided. The big advantage of infrared is that it does not require oxygen to function and can read in oxygen-deficient atmospheres. The device also is not affected by temperature, nor is it easily poisoned by high exposures. Disadvantages include cost and many cross-sensitivities. This is another sensor that should not be purchased as a stand-alone unit, but should be considered after the purchase of a catalytic bead or Wheatstone bridge sensor.

Toxic Gas Monitors

This section will describe the sensors that are commonly used in three-, four-, or five-gas units (LEL, O_2, toxic/toxic/toxic) and are usually used to measure carbon monoxide (CO) and hydrogen sulfide (H_2S). Toxic sensors are available in a variety of gases, including chlorine, sulfur dioxide, hydrogen chloride, hydrogen cyanide, nitrogen dioxide, and many others. The most common unit sold today by far is a four-gas unit that measures LEL, O_2, CO, and H_2S; its popularity is a direct result of the confined space regulation issued by OSHA. Many people understand the first three sensors but don't understand why H_2S was chosen as the common fourth gas. OSHA does not specify H_2S, but it is a common gas found in confined spaces, and most confined space entries (and consequently most deaths and injuries) take place in sewers and manholes. H_2S is commonly found in sewers and any place else organic materials are rotting. Thus, a common recommendation is to stick with H₂S as the fourth or fifth sensor unless a department has a specialized need for some other gas detection.

Response teams must consider carefully which toxic sensors they most require. They must weigh the cost of making this decision. There are other more cost-effective methods of detecting toxic gases such as chlorine or sulfur dioxide. The average cost for a sensor is $400 to $500, and the equipment is usually guaranteed for one year. The calibration gas is $300 to $400 and is only good for six months before it expires. The sensitivity of these detectors is another issue: If you take an instrument into an atmosphere with more than 20 ppm chlorine, you will ruin the unit.

Most toxic sensors are electrochemical sensors with electrodes (two or more) and a chemical mixture sealed in a sensor housing. The gases pass over the sensor, causing a chemical reaction within the sensor and an electrical charge, which causes a readout to be displayed. All toxic sensors display in parts per million. Some toxic sensors use metal oxide technology and react in the same fashion as an MOS.

Other Detectors

In most cases these other detectors will be used by a HAZMAT team, but the first responder should be aware of the devices and their capabilities. In areas without a HAZMAT team or where the team has unusually long response times, the first responder may want to consider the purchase of these items. As the other instruments are complicated, so are these instruments. First responders are capable of beginning the detection process, but in almost all cases further detection capability is required.

PHOTOIONIZATION DETECTORS (PID)

The **photoionization detector (PID)** is sometimes referred to as a *total vapor survey instrument;* an example is shown in **Figure 6-22.** Because of their ability to detect a wide variety of gases in small amounts, PIDs are becoming essential tools of HAZMAT response teams. The PID will not indicate what materials are present, much the same as the CGI will not identify the specific material present. Used as a general survey instrument, the PID can alert responders to potential areas of concern and possible leaks/contamination. These instruments look for toxic materials in the air and are essential to determining exposures to many toxic materials. The combustible gas detector will start reading toxic materials in a range of 50 to 500 ppm, which for some gases is extremely high and could

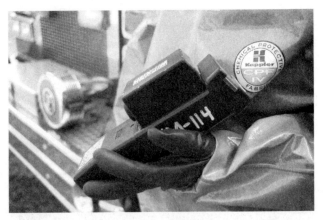

Figure 6-22 The photoionization detector (PID) is used by HAZMAT teams to detect toxic gases in the air. It is a very sensitive instrument and measures small amounts of gases in the air.

pose immediate risks for a responder. The PID starts to read at levels near 0.1 ppm, which is very sensitive. Because of their sensitive nature, they can detect small amounts of hydrocarbons in the soil. As sick building calls are on the increase, the PID is a valuable tool in identifying possible hot spots within the building.

COLORIMETRIC SAMPLING

Colorimetric tubes are used for the detection of known and unknown vapors. As sick building calls become more frequent, the use of colorimetric tubes by HAZMAT teams has greatly increased. Colorimetric sampling consists of taking a glass tube filled with a reagent, usually a powder or crystal, that is placed into a pumping mechanism that causes air to pass over the reagent. A colorimetric sampling device is shown in **Figure 6-23.** If the gas reacts with the reagent, a color change should occur, indicating a response to the gas sample. Detection tubes are made for a wide variety of gases and generally follow the chemical family lines (i.e., hydrocarbons, halogenated hydrocarbons, acid gases, amines, etc.). Although the tubes may be marked for a specific gas, they usually have cross-sensitivities, or react to other materials, which at times is the most valuable aspect of colorimetric sampling.

A new system that has been developed involves the use of bar-coded sampling chips. This system, called the **chip measurement system (CMS),** involves the insertion of a sampling chip into a pump, which is shown in **Figure 6-24.** The pump recognizes the chip in use and provides the correct amount of sample through the reagent. Using optics and a light transfer system, the reflective measurement provides an accurate reading of the gas that may be present.

Figure 6-23 A standard colorimetric tube is a glass tube with a crystalline material that changes color when a certain type of gas is present. These sensors can be used to identify unknown gases and can provide a level of a known gas.

Figure 6-24 The chip measurement system (CMS) uses a chip to analyze a gas and provides a digital readout. Standard colorimetric sampling requires an interpretation of the color change and the length of the change. The CMS system takes the guesswork out of the system by providing the reading in ppm.

Other instruments used by a HAZMAT team include pH detection, flame-ionization detectors, warfare agent detection, mace/pepper spray detection, and characterization sample kits. The detection of unknown materials is complicated even for some HAZMAT teams. When first responders are confronted with a possible chemical release—no matter how insignificant they think it is—they should request or as a minimum consult with their closest HAZMAT team.

CARBON MONOXIDE INCIDENTS

Among calls on the increase for the fire service is the response to carbon monoxide detector alarms. In 1995 the city of Chicago ran several thousand carbon monoxide detector alarms in one day, due to an inversion, which kept smog, pollution, and carbon monoxide at a low elevation within the city. As carbon monoxide is colorless, odorless, and very toxic, it is important that first responders understand the characteristics of carbon monoxide and how home detectors

work. As it is with other chemicals, carbon monoxide (CO) can be an acute or chronic toxicity hazard. It is only acutely toxic at high levels, typically at levels in excess of 100 ppm. At levels less than 100 ppm the hazard comes from a chronic exposure. A chronic exposure to carbon monoxide can be hazardous since many people are exposed to CO in the home. The time people spend in their homes constitutes a considerable exposure. CO can only be detected by a specialized monitor. In extremely high concentrations it can be explosive. Exposure to CO causes flu-like symptoms, headache, nausea, dizziness, confusion, and irritability. Exposure to high levels can cause vomiting, chest pain, shortness of breath, loss of consciousness, brain damage, and death.

When exposed to levels of CO in a house, the residents may not be exhibiting any signs or symptoms. If CO is detected, they should seek medical attention because CO poisoning signs and symptoms may be delayed for twenty-four to forty-eight hours. Levels over 100 ppm are extremely dangerous, and residents should be medically evaluated. Women who are pregnant should be advised to be checked, as there have been fetal fatalities at 9ppm CO. Monitoring with a CO detector is essential to determine the possible exposure to CO. When people show only minor effects of CO poisoning and would normally be transported to the closest hospital, their home should be checked for CO levels. If high levels are found using a monitor, then alternative treatment, such as a **hyperbaric chamber,** may be the best treatment and should not be delayed. Many times a Pulse-Ox (oxygen saturation monitor) will be used incorrectly to determine a patient's O$_2$ level. Patients who have been exposed to CO will cause a Pulse-Ox to read 100 percent as the monitor reads the oxygen molecule in CO as being O$_2$. The elderly, children, or women who are pregnant are especially susceptible to CO and may have had a serious exposure without showing any effects.

If the first arriving units do not have a CO monitor and victims may be remaining in the residence, personnel are to don functioning SCBA when searching the residence. After determining no victims are present, crews are to ensure that the house is closed up and then wait outside for a CO monitor. Do not enter an area with a CO detector that is activated without the use of SCBA. If crews find levels that exceed 35 ppm according to the monitor, they should use SCBA to continue the investigation. Crews should be suspicious when responding to reports of an unconscious person or of "several persons down." They should not enter an area without

SCBA if it is possible that CO (or other toxic gases) may be present. An air monitor will ensure responder safety from the gases for which it samples.

It is possible that people may be found unconscious due to a natural gas leak, but this is very unlikely. Natural gas, which has a distinctive odorant added to it, is nontoxic and will only asphyxiate a person by pushing oxygen out of an area. The only sign of this exposure is unconsciousness or death; any flu-like symptoms are due to CO poisoning. If the level of natural gas is high enough to cause unconsciousness, a very severe explosion hazard is present, and in reality an explosion would be imminent.

It is possible that standard fire department air monitors will not pick up any CO. This is because the CO detectors purchased for the homes are made to detect small amounts of CO over a long period of time, and fire department detectors are instant readings that only pick up 1 ppm or more. The fact that responders do not pick up any readings does not mean the residents' detector is defective. Some factors that account for this discrepancy are low amounts of CO or a momentary high level that activated the alarm and dissipated prior to responders' arrival. The amount of time the residence is open dramatically affects readings. Crews are reminded to keep the residence as closed as possible so that

TYPES OF CO SENSING TECHNOLOGY

Biomimetic employs a gel-like material designed to operate in the same fashion as the human body when exposed to CO. This sensor is prone to false alarms, as it can never reset itself, unless it is placed in an environment free of CO, which in most homes is impossible. The sensor may need twenty-four to forty-eight hours to clear itself after an exposure to CO. The actual concentration of CO at the time the detector sounds may be low, but the exposure may have been enough to send the detector over the alarm threshold. If responding to an incident in which one of these detectors has activated, it should be placed in a CO-free environment until it clears. In meantime, residents should rely on an alternate CO monitor.

The metal oxide detector is similar to sensors used in combustible gas detection but is designed to read carbon monoxide. How successful this design is in only reading CO is subject to debate. In some respects, the metal oxide monitor is superior to the biomimetic sensor, but it does have some cross-sensitivities and will react to other gases. For example, it is possible for this type of sensor to sound an

alarm for propane. Responders using only a CO instrument may find themselves walking into a flammable atmosphere. Metal oxide detectors can usually be identified by the use of a power cord as the sensor requires a lot of energy. In most cases these sensors provide a digital readout. Once activated this sensor needs some time to clear itself, usually less than twenty-four hours.

An electrochemical sensor, which may also be referred to as instant detection and response (IDR), is the same type of electrochemical sensor that is in your three-, four-, or five-gas instrument. It has a sensor housing with two charged poles in a chemical slurry. When CO goes across the sensor, it causes a chemical reaction, which changes the resistance within the housing. If the amount is high enough, it will cause an alarm. This sensor provides an instant reading of CO and does not require a CO buildup to activate. It has an internal mechanism that checks the sensor to make sure it is functioning, which is a unique feature. Out of the three types of residential detectors, based on sensor technology, the electrochemical monitor would provide the best sensing capability.

the air monitor has a chance to establish the level of CO. As a reminder any time you respond to unknown odors or sick building calls, it is important to remove people from the building and keep it closed up. As the amounts of toxic gases in sick building incidents are usually small, keeping them contained is very important. You cannot treat patients of toxic gas exposure unless you have a clue as to the source. For the patients' long-term health and your own well-being, you need to obtain quick reliable gas samples.

In some parts of the country CO detectors are required just as smoke detectors are. Performance varies from brand to brand. The three basic sensing technologies are biomimetic, metal oxide, and electrochemical, each with advantages and disadvantages. Location, weather conditions, and the type of sensor determine what types of readings can be expected from a particular brand of detector.

Common sources of carbon monoxide include:

- Furnaces (oil and gas)
- Hot water heaters (oil and gas)
- fireplaces (wood, coal, and gas)
- Kerosene heaters (or other fueled heaters)
- Gasoline engines running inside basements or garages
- Barbecue grills burning near the residence (garage or porch)
- Faulty flues or exhaust pipes

SUMMARY

The use of defensive product control methods is a key component for the protection of the community and the environment. In most cases first responders have the equipment necessary to handle these tasks. With some modification or adaptation first responders can accomplish the control of many spills. The limiting of spills will mitigate the incident sooner and prevent its spread. If first responders cannot stockpile the necessary equipment, then contacts should be made for those facilities that may have the materials. Under the requirements of the Oil Pollution Act (OPA) of 1990, certain facilities are required to maintain stockpiles of emergency equipment. First responders are also becoming more involved with air monitoring and are becoming more aware of the hazardous chemicals present. When using air monitors, first responders are reminded that they are not all encompassing and require training and experience to use properly. They require testing, maintenance, and regular repair and cannot be expected to function

properly without this upkeep. To determine true safety first responders must use a range of instruments, something which is usually above the first responder level. When in doubt, responders should consult with their local HAZMAT team.

KEY TERMS

Absorption   Defensive method of controlling a spill by applying a material that absorbs the spilled chemical.

Biomimetic   Form of a gas sensor that is used to determine levels of carbon monoxide. Typically used in residential CO detectors, it closely recreates the body's reaction to CO and activates an alarm.

Bump test   Used to determine if an air monitor is working; will start alarm if a toxic gas is present. This quick check ensures the instrument responds to a sample of gas.

Calibration   Setting the air monitor to read correctly. To calibrate a monitor, the user exposes it to a known quantity of gas and makes sure it reads the values correctly.

Catalytic bead   The most common type of combustible gas sensor, which uses two heated beads of metal to detect the presence of flammable gases.

Chip measuring system (CMS)   Form of colorimetric air sampling in which the gas sample passes through a tube. If the correct color change occurs, the monitor interprets the amount of change and indicates the gas level on a LCD screen.

Colorimetric tubes   Used to detect both known and unknown types of gases. A form of detection for toxic gases, which change color in the presence of a toxic gas.

Combustible gas indicators   Detection devices used to measure flammable gases in air.

Damming   Stopping a body of water, which at the same time stops the spread of the spilled material.

Diking   Defensive method of stopping a spill; a common dike is constructed of dirt or sand and is used to hold a spilled product. In some cases, in a facility a dike may be preconstructed, such as around a tank farm.

Dilution   Adding a material to a spill to make it less hazardous; in most cases water is used to dilute a spilled material, although other chemicals could be used.

Diverting   Using materials that can move products around or away from an area; for example, several scoops of dirt can be used to divert a running spill around a storm drain.

Hyperbaric chamber   Chamber used to treat scuba divers who ascend too quickly and need extra oxygen to survive. The chamber, which recreates the high pressure atmosphere of diving and forces oxygen into the body, is also used to treat carbon monoxide poisonings and smoke inhalation.

Infrared sensor Monitor that uses infrared light to determine the presence of flammable gases. Light is emitted in the sensor housing and the gas passes through the light; if it is flammable, the sensor indicates the presence of the gas.

Metal oxide sensor (MOS) Metal oxide coated LEL sensor that is heated to detect the presence of flammable and other gases.

Multi-gas detector Air monitor that measures oxygen levels, explosive (flammable) levels, and one or two toxic gases, such as carbon monoxide or hydrogen sulfide.

Photoionization detector (PID) Air monitoring device used by HAZMAT teams to determine the amount of toxic materials in the air.

Remote shutoffs Valves that can be used to shut off the flow of a chemical; the term *remote* is used to denote valves that are away from the spill.

Retention Digging a hole to collect a spill; can be used to contain a running spill or collect a spill from the water.

Vapor dispersion Intentional movement of vapors to another area, usually by the use of master streams or hoselines.

Vapor suppression Defensive strategy to minimize the production of vapors. For example, flammable liquids produce hazardous vapors that could ignite; firefighting foams are used to suppress the vapors and not allow them to move by forming a barrier on top of the spill.

Wheatstone bridge sensor Type of combustible gas sensor that uses a heated coil of wire to detect the presence of flammable gases.

REVIEW QUESTIONS

1. What type of dam would be required for a fuel spill in which the fuel has a specific gravity of less than 1?

2. What type of dam is required for a chemical spill in which the material has a specific gravity of greater than 1?

3. What are two key items needed to construct an underflow or overflow dam?

4. Who should be consulted prior to using vapor dispersion techniques?

5. What is the normal configuration for a multi-gas detector that is used by first responders?

6. If you have a detector that is calibrated for methane, and you respond to a propane release, will the instrument detect propane? If yes, what do you need to know about propane and the detection device?

7. How often should detectors be calibrated?

8. In regard to safety and the ability to detect various toxic gases, how would you rate the use of a multi-gas detector (LEL, O_2, CO, H_2S)?

   a. Provides a high margin of safety

   b. Provides a low margin of safety

Suggested Readings

Hawley, Chris, *Air Monitoring and Detection Devices*, Delmar, a Division of Thompson Learning, Albany, NY 2001

Lesak, David, *Hazardous Materials Strategies and Tactics*, Prentice Hall, New Jersey, 1998.

Noll, Gregory, Michael Hildebrand, and James Yvorra, *Hazardous Materials Managing the Incident*, Fire Protection Publications, Oklahoma University, 1995.

Schnepp, Rob, and Paul Gantt, *Hazardous Materials: Regulations, Response, and Site Operations*, Delmar, a division of Thomson Learning, Albany, NY, 1999.

Smeby, L. Charles (editor), *Hazardous Materials Response Handbook*, 3rd ed., National Fire Protection Association, 1997.

TERRORISM AWARENESS

STREET STORY

In light of the terrorist events that have befallen New York City, Oklahoma City, and Tokyo, and certainly encouraged by the recent rash of anthrax "scares" in Los Angeles and numerous other cities in December 1998, I have been asked several times, "What is Sacramento doing to prepare for a potential act of terrorism that has yet to visit us?" I pause for a moment and think. Can I imagine the chaos and confusion generated among emergency response agencies during such an event? This query implies that I have had no such experience. What in Sacramento could ever be so attractive to terrorists that they would be tempted to plan an event to kill, maim, and disrupt?

The time was 2:20 P.M. The date was April 24, 1995. I remember it well. I was temporarily at our Division of Training. In fact, I was talking long distance to one of our US & R team leaders, Sacramento CA-TF#7, having been on scene in Oklahoma City for four days. So many people had been reported killed. I had been ordering rescue equipment for them and acting as a general liaison since their departure for Oklahoma City. As a fire officer, one tends not to pay close attention to background radio traffic on a daily basis. Only key bits and pieces of words and phrases of various dispatches stay in one's mind. The same is true with typed messages. With phone in hand, I was writing down lines of equipment part numbers, not paying much attention to a dispatch "active status" screen being displayed in front of me. Nor did I hear the last call, a report of smoke coming from a brick office building at a downtown Sacramento address. I was more interested in when the vendors were going to return my flurry of phone calls. Then suddenly, I sat straight up and said, "Oh my God." My brain began to tuning to something it had just heard. On the other end of the phone in Oklahoma City, the team leader was asking me what was wrong, and I was trying to focus on what I was hearing.

"This is Engine 14. I repeat, this is to confirm a bomb has exploded at this address. There is one fatality." Our dispatch center acknowledged the report. I turned and scanned the active status screen on the CAD terminal. Displayed as a working incident was Thirteenth Street and I Street. I froze. It was the address of the Sacramento Fire Department Headquarters. I do not remember saying goodbye to the person in Oklahoma City, nor do I remember hanging up the phone. I was thinking, why would anyone want to blow up my headquarters? Who was killed? I was on scene in less than three minutes. Confusion? Chaos? Grave concern? Redundant questions. It was typified everywhere among fire, law, and federal agents who were arriving. The responsibility of PIO for this incident was mine for the next eight hours.

The trial took six months. The federal jury found Theodore Kaczynski, the Unabomber, guilty of murder and he was sentenced to life in prison. It was the second murder perpetrated by Kaczynski in Sacramento and the Unabomber's eighteenth and last act of terrorism. Today images of chaos, panic, and anxiety still flash through my mind as I reflect on that scene. So when did Sacramento begin taking terrorism events for real and begin planning in earnest to improve training and management skills? Literally, the next day.

—Street Story by Jan R. Dunbar, Division Chief, Special Operations, Sacramento Fire Department, Sacramento, California

OBJECTIVES

After completing this chapter the reader should be able to identify or explain:

- Potential target locations (A)
- Indicators of potential terrorist activity (A)
- Incident actions to be taken at a terrorist attack (A)
- Additional hazards at a terrorist attack (O)
- Other specialized resources to assist with a terrorist attack (O+)
- Methods of requesting federal assistance (0+)
- Common agents that may be used in a terrorist attack. (A)

INTRODUCTION

It is unfortunate that in this text that a chapter on terrorism needs to be included. Until recent times this would not have been necessary. Although we have all seen terrorism on the evening news, it was typically in places such as Northern Ireland, Beirut, Israel, or another country. The United States had for the most part remained immune to the reign of a terrorist. This changed in February 1993 when the World Trade Center was bombed in New York City (see **Figure 7-2**). Even when this bombing occurred, the responders did not pay much attention, as the bombing was considered as big city or as an incident that could only happen in New York City. Many HAZMAT instructors tried to inform their students as to the ease of making such a bomb, but for the most part the warnings fell on deaf ears. The people who were found to be responsible for the bombing were controlled by an ideology from outside the United States, and so the thought was that it was an isolated international terrorism attack. When the Alfred P. Murrah Federal Building in Oklahoma City was devastated by a bomb on April 19, 1995, the United States emergency responders took notice. The attack on the Murrah building, which is shown in **Figure 7-3,** was perpetrated by Timothy McVeigh and Terry Nichols, who did not have ties to another country and are natural-born citizens. It was perceived as an attack on our country from one of our own, and not some unknown foreign nation. It brought terrorism home for the public and emergency responders alike.

Figure 7-2 A van was used to carry explosives to the underground parking garage of the World Trade Center. Six people were killed and more than 1,000 were injured. There were 50,000 people in the building, and the goal of the terrorist was to collapse the building into the adjacent tower.

Figure 7-3 A truck bomb caused the devastation in the Oklahoma City bombing, killing 167 people and injuring 759. The damage extended several blocks in each direction, and 300 buildings were damaged. Fatalities occurred in fourteen separate buildings.

Regardless of its origin, the potential for terrorism is here in this country and remains in our thoughts as we respond to any incident. Other incidents are occurring on a regular basis, although they do not fit the exact definition of terrorism, involving the use of large caliber weapons or a large number of weapons. The Federal Bureau of Investigation (FBI) in 1997 investigated 72 potential incidents of terrorism, and in 1998 they investigated 146 cases. In 1999 they had investigated more than 256 cases, with a good majority of the cases involving anthrax hoaxes. Other than hoaxes, most of these crimes are aginst individuals, attempted murders or murders for hire. Anthrax hoaxes have lost a lot of thier impact, and will be replaced by another hoax, most likely are radioactive events. These numbers demonstrate the dramatic increase in these types of events. Many criminals are protecting themselves with body armor and fortified vehicles. Many crimes such as bank robberies involve the use of explosives and sophisticated weaponry. Street violence, often associated with gangs, is becoming increasingly violent. When people are killed or injured, the fire service is called into action.

Crimes such as assaults and murders are even increasing within our schools. In April 1999 fifteen people (one teacher and fourteen students, including the two attackers) died in a suicide attack by two students on Columbine High School in Littleton, Colorado. Each year there has been an increase in the amount of violence in our schools. The EMS and fire services are placed into action in these events and in many cases arrive first. In recent times there have been deadly riots in Los Angeles and in St. Petersburg, Florida both of which had a significant impact on the fire service. Even the workplace is becoming an increasingly dangerous place for emergency responders. In April 1996 a firefighter in Jackson, Mississippi, entered the headquarters fire station intent on killing the fire chief; several people were killed and the emotional damage was widespread.

Crimes are becoming more violent and it is perceived that this trend will continue to intensify. Emergency responders are immediately placed into a dangerous situation and can get caught, literally and figuratively, in the cross fire. As a result of the World Trade Center bombing, the Murrah building bombing, and other incidents we will discuss, our procedures for some incidents have been forever changed. Since these two large attacks there have been two other bombings that are fairly well known to the fire service, the Atlanta Olympic park bombing in 1996 and the Atlanta abortion clinic bombing in 1997. That particular abortion clinic bombing is significant because a secondary device was used. It is believed that it was strategically placed with the sole intention of harming responders. The device was placed near the location where the IC post was set up. Luckily only minor injuries were received by the responders who were near the blast. In 1998 in another bombing at a Birmingham, Alabama, abortion clinic, an off-duty police officer employed as a security guard was killed by a device that was more accurately placed to target the responders. This secondary device was activated by the person responsible as he watched the victim approach the radio-controlled device.

Among major terrorist attacks and related incidents in the United States in recent years:

■ A series of bombings in the 1980s targeted primarily at the Internal Revenue Service included an attempt at a chemical bomb using ammonia and bleach. Another attempt included the use of a hot water heater as a very large pipe bomb, but the vehicle carrying the bomb caught fire while the terrorist was driving it.

■ In 1993 a bomb in a van killed six people at the World Trade Center.

■ A man used his employer's Nuclear Regulatory Commission (NRC) license to order several radioactive sources in 1996. The purpose of the purchase was not determined, but foul play was suspected.

■ In 1996 members of a paramilitary group who had access to preplanning information obtained through the fire department were arrested when they attempted to blow up a Department of Justice complex in West Virginia.

■ A man was arrested in 1995 after manufacturing **ricin,** which is an extremely deadly toxin, with the intention of killing a rival.

■ In 1996 and 1997 the Atlanta area was besieged with bombings, including one at the Olympic Park, which killed one person, and another at an abortion clinic in Birmingham, Alabama, in which a booby-trapped device killed an off-duty police officer and wounded a nurse.

■ In April 1995 the Alfred P. Murrah Federal building in Oklahoma City was the target of a truck bomb. Several hundred people were injured and 186 people were killed in the blast.

■ Members of a militia group known as the Patriots Council were arrested in 1995 for the manufacture of ricin in an attempted assassination of a U.S. marshal.

■ A train was derailed near Hyder, Arizona, in 1995, and a terrorist group claimed responsibility, although no one has been charged in the attack. One person was killed and the derailment seriously injured twelve people.

■ The Anaheim (California) Fire Department was alerted to a potential sarin nerve agent attack at

Disneyland in April 1995. The chief was notified at midnight to be at the command post four hours later. Police agencies, federal law enforcement officers, and the military had known about the threat for five days. Up until notifying the fire chief and assembling the resources prior to the event, no other planning had occurred. The fire department was placed in charge of the incident, and quickly developed a plan of action. Disneyland did not close and 7,000 to 40,000 people visited the park during the threat period. Luckily, the threat never materialized.

■ Two men were arrested in 1998 on suspicion of possessing **anthrax.** They were later released, since it was determined that they only possessed a possible anthrax vaccine. One of the men had previously been arrested in 1995 for the possession of a plague microbe, which he had ordered through the mail. Both were alleged members of an antigovernment group.

■ In Charlotte, North Carolina, a man with an explosive device held hostages in a government building in 1998. The device was thought to contain some type of chemical agent. It was later determined that the filling agent was harmless, although the explosive was live.

■ Abortion clinics in South Florida, New Orleans, and Houston were affected by attacks using **butyric acid** in 1998.

■ In 1998 a letter with a powder reported to be anthrax was found in the courthouse in Wichita, Kansas. More than 200 people were evacuated, while some were decontaminated and treated at the local hospital. The substance was later found to be harmless.

■ A pickup truck rammed the courthouse in Lafayette, Indiana, in 1998. The bed of the pickup held flammable and combustible materials and several explosives.

■ In 1999 two students who were armed with an array of guns and explosives attacked their own high school in Littleton, Colorado. The suicide attack resulted in 15 deaths and spurred a rash of bomb hoaxes throughout the country.

■ In early 1999 the FBI investigated hundreds of anthrax hoaxes, none of which involved the actual use of the biological agent. Abortion clinics were the targets in most of the cases.

In this chapter we will provide some information about the type of terrorism agents that currently exist along with some possible scenarios. There exists some concern within the emergency services that we should not publish this type of information. Taking this into account, it is understood that all of

this information, plus additional reference sources, are taken from the Internet, which is readily available in the public domain, in training programs, and in other accessible material throughout the United States. Most of the exact recipes and how-to instructions for these and many other devices are easily obtained through printed texts and the Internet. We are not providing any information that is not readily accessible elsewhere. Since we are able to acquire this data in such an easy, normal, and legal fashion, rest assured that terrorists already have it. Responders should be aware of the various chemicals and devices that someone may design to kill the first responders and others in your community. The best defense is a broad-based education, with the challenge of staying one step ahead of the terrorist. By staying informed as to how terrorists may operate, you can be more alert to potentially fatal circumstances.

TYPES OF TERRORISM

The types of terrorism are divided into two distinct sources: foreign based and domestic. Until the Murrah building attack the fear of terrorism was directed at a foreign source. Most Americans believed that any terrorist attack would be from another country. To be deemed foreign based, the motivation or supervision comes from another country. Domestic terrorism originates from within the United States and is not influenced by any foreign party. The perception is that our greatest threat from terrorism is international in nature, but the opposite is proving to be more accurate. From the list provided earlier in this chapter and from other compiled statistics, it is evident that the largest percentage of terrorism has been domestic in origin.

NOTE The FBI defines terrorism as a violent act or an act dangerous to human life, in violation of the criminal laws of the United States or any segment to intimidate or coerce a government, the civilian population, or any segment thereof, in furtherance of political or social objectives.

The key to this definition is the intimidation of the government or the civilian population. In 1984, a religious group trying to influence the local political process sprayed salmonella on a fast-food restaurant salad bar; the attack rendered more than 715 people ill. This attack was originally not considered an act of terrorism, as the health department initially suggested it was food poisoning. Not until more than a month later did the FBI learn that it was an intentional act. This information was discovered while investigating another crime involving the same group.

The Tokyo subway sarin attack was an attempt to destroy a good portion of the police department, which was trying to raid the terrorist compound. Although the main topic of this chapter is terrorism, that label applies to only a minority of the types of incidents to which firefighters respond.

SAFETY Many incidents may not fit the exact terrorism definition, but the hazards from a pipe bomb are the same regardless of the motivation of its builder.

Many responses that could have been routinely handled in the past must now be treated much differently, and responders must always be on their guard for terrorist-style devices or potential terrorism acts.

The terrorist's motivation is to produce fear, which may be aimed at the general public or the government. This goal can be accomplished by large scale actions such as the bombing of the World Trade Center or the Murrah building. It may even be on a greater scale, like the sarin attack in Tokyo. A terrorist can also incite this level of fear by planting the threat of terrorism or by devising a hoax. The latter scenario is the most likely, and can be very difficult to handle from an emergency service perspective. Another consideration, as we have seen in England and Ireland, is the disturbing trend to view emergency service personnel as targets. One theory currently under examination is that the second explosion at one of the Atlanta abortion clinics was aimed at the emergency responders.

SAFETY Responders must always be alert and aware of the potential for a secondary device.

When responding to a possible incident of terrorism involving **weapons of mass destruction (WMD),** there is always the likelihood that the incident could be a hoax. When dealing with potential explosive incidents the statistics show that there are thousands of explosions carried out each year. Although training for a variety of emergencies is crucial to success, responders must determine their most likely event. Most of the explosive incidents involved small devices and are not nearly on the scale of the World Trade Center or Murrah building. No matter the material or the device, several criteria must be considered when determining the validity of a terrorism threat: educational level, availability of raw materials, production equipment, dissemination, and motivation.

POTENTIAL TARGETS

Potential targets exist throughout every community in the nation. Terrorists may select commercial buildings, high-rise buildings, even residential homes. Although some incidents do not fit the definition of terrorism, the materials used are the same as in a terrorist situation. Whether the objective is terrorism or some other type of crime, the danger to the responder is the same. When considering terrorism, potential targets may be grouped into several categories: public assembly, such as the area shown in **Figure 7-4,** federal, state, and local public buildings; mass transit systems, high economic impact areas; telecommunication facilities; and historical or symbolic locations.

While obviously not an exclusive list, some buildings should be of concern, including FBI, ATF, and IRS offices; military installations; Social Security buildings; transportation areas; city or county buildings, including fire and police stations; abortion clinics; Planned Parenthood offices; fur stores; laboratories; colleges; cosmetic production/testing facilities; banks; utility buildings; churches; and chemical storage facilities. Transportation facilities such as airports and train, bus, or subway stations are high on the potential list of targets. Given the number of people who may be potential targets and the relative ease of escape, these areas are open for attack. In the southeast United States, several churches were damaged or destroyed in arson fires in the 1990s, and in some cases explosive devices were used. Many abortion clinics have been the subject of bombings, attacks, and other threats. Any incident in or near one of these facilities should be approached with caution. Responders should know the location of these facilities in their jurisdiction. Preplans for these facilities should be developed by the company officers, but committing these plans to paper is not recommended to avoid leaving a blueprint for a terrorist. As the battle between groups against abortion and for reproductive choice continues to heat up, incidents at these facilities can only be expected to rise, with emergency responders caught in the cross fire. Family planning clinics are not the only places targeted in this political struggle. Several anti-abortion groups have posted the home addresses of family planning

ASSESSING TERRORIST THREATS

The investigation to assess the credibility of a threat has five elements. If the person known or thought to be responsible can be determined to have several of these capabilities, the credibility factor increases. The first of the five questions involves the potential terrorists' ability to make a device or agent. Other than explosives or ricin, this knowledge is very difficult to attain. To make a biological pathogen agent, in most cases one needs an advanced knowledge in biology. On the other hand, ricin is a biological toxin that does not require any advanced knowledge to produce. Some threats with letters or packages have misstated the origin of the material. The terrorist might call anthrax a virus, for example, or misspell the agent's name. It follows that if terrorists do not know the true origin of the material or cannot spell it correctly, they probably don't have the education necessary to make the material. However, this assumption does not take into account a person who may purchase the material.

Another element is a person's ability to obtain the raw materials necessary to make the agents. Many of the materials necessary to make chemical warfare agents are banned for sale, and others appear on hot lists, which means they are only sold to legitimate businesses. This would not preclude someone from buying the raw materials on the black market or stealing them from a legitimate business.

The third element is the ability to manufacture the devices or machinery required to make the device or agent. To manufacture chemical warfare agents requires the use of a reactor vessel, which to produce less than a gallon requires about a 10-by-10-foot space. Some agents could be produced in a bathtub using backyard chemistry, but these are not the high-end agents that attract so much attention. When people attempt to make agents in less than ideal conditions, a frequent result is that they end up killing themselves during the production process. Many criminals do not take the time to follow standard industrial safety precautions.

A crucial element is the ability to disseminate these agents. The military conducted many tests on chemical and biological warfare agents; and although it has some good methods of dissemination, it still lacks a 100 percent effective method. The Aum Shinrikyo cult in Japan is a perfect example. This group, with millions of dollars in assets, full chemical and biological lab, and production facilities, employed the services of 235 scientists to develop and manufacture chemical and biological agents. The Aum Shinrikyo abandoned its biological weapons program after a full-scale release of anthrax failed. The group used sarin twice, first in Matsumoto where seven people were killed and 200 injured. The dissemination method used in the Matsumoto attack was much more effective than the one in the Tokyo subway. If the group had used the same dissemination method, it would have greatly altered the course of events. Aum Shinrikyo would have been limited by the amount of agent that could have been produced in a short period of time. It is commonly thought that, despite its relatively vast resources, Aum Shinrikyo was only able to make a small amount of agent.

A final and crucial consideration is whether the person or group has the motivation to pull off the attack. The intentional killing of one person takes considerable motivation, and the motivation required to kill hundreds requires exponentially more.

No one should dismiss the potential for an attack, but the point is that it takes considerable education, raw materials, manufacturing, and dissemination ability to pull off a chemical or biological WMD attack. On the other hand, explosives are easy to manufacture and do not require much education. They do not require anything more than simple tools, and the materials required are easy to assemble. For this reason explosives are used in the majority of attacks, often with great destructive power. In many cases the terrorist can be successful because of the hysteria associated with a potential terrorism incident. A balance must be struck between a cautious approach and one that does not allow the terrorist to win by crippling a community and causing hysteria.

physicians on the Internet, with the alleged intent of promoting violence against these doctors.

Many of the potential targets of some groups have not been buildings at all but events where large numbers of people are present. The Atlanta Olympic Park bombing is an example. Other scenarios involve sports stadiums, public assembly locations, transportation hubs, fairs, and festivals (see **Figure 7-5**). First responders should have some preplans for these types of locations. One possible scenario for a sta-

dium devised by Captain Richard Brooks of the Baltimore County Fire Department describes the first-in medic unit to a stadium where in section 70 there are 40 people projectile vomiting. After five minutes 200 people are projectile vomiting, and as time goes on the number increases. What happens to the responders when confronted with a situation of this nature? How many responders would be affected by this massive amount of afflicted people? How many responders would be needed to handle this incident?

Figure 7-4 Any location is a potential target for a terrorist. Any location where large numbers of people are present, such as a mall or sports event, are prime targets. Many of the arrests that are related to terrorism are in small towns and rural areas. The assumption that terrorism is limited to urban areas is proved incorrect by arrest statistics.

Figure 7-5 Other than special events, the most common location where large numbers of people are together are sporting events. At this stadium, if an incident were to occur, more than 50,000 people can become part of the incident.

This could be accomplished by putting syrup of Ipecac™ in the ketchup container beside one of the hot dog vendors, an easy task. Imagine the hysteria if a note was found stating that a biological agent was distributed in that section. What impact would that have on the remaining 50,000 people in that stadium if the threat got out? Planners and responders involved on the national level to develop response profiles to terrorism are grappling with how to plan for incidents involving 100 people, 1,000 people, 10,000 people, and 50,000 people. Terrorism incidents can very quickly overwhelm the responders and their whole emergency response system.

When dignitaries visit locations, typically there is a lengthy planning process, and the fire department should be involved. When the Pope visited Baltimore in 1997, the planning process took more than eight months. Planning for such a large event takes the cooperation of local, state, and federal agencies. Even when dignitaries visit locations such as New York City or Chicago, there is some advance planning. Other events such as political conventions also bring the potential for an incident. When one of these events come to your community, you need to be prepared for not only the people coming to the event but also the massive federal response which may be pre-positioned. For many of the events there are usually whole task forces of federal resources hidden away just in case of an incident.

Certain dates have significance to several militant groups. April 19 is significant as the anniversary of the ATF storming the Waco, Texas, compound that housed the Branch Davidians, a group thought to be a militia group. April 19 was also, not coincidentally,

the date of the Oklahoma City bombing. The significance of April 19 goes back to the Revolutionary War. Another symbolic date is the anniversary of Roe vs. Wade[1] on January 22, which harbors the potential for anti-abortion groups to strike. Within the United States there are a number of militia, patriot, or constitutionalist groups; together they are suspected to have members in forty states. Membership counts vary from fifty in some states to several thousand members in other states. Groups of concern include anarchist groups and white supremacy groups such as the Ku Klux Klan, skinheads, and neo-Nazi groups, including the Aryan Nation. Other groups suspected of activity or thought to have terrorism potential include patriots, militia, constitutionalists, and tax protesters. To learn which groups are active, it is relatively simple to use a web browser and search the Internet for many of these groups; most have web sites.

INDICATORS OF TERRORISM

An explosion or explosive device is the most common tool of the terrorist, and police across the country have made several arrests of people making and/or storing large quantities of explosives.

CAUTION According to an FBI source, more than 93 percent of terrorist incidents use explosives as the weapon of choice.

The most common device is a pipe bomb, such as the one shown in **Figure 7-6.** In any incident

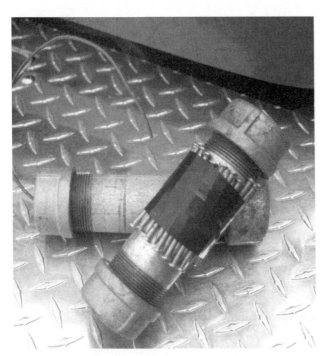

Figure 7-6 The most common explosive device is a pipe bomb, and it is very effective. It is a very dangerous device, not only for responders but for the builder as well. The pipe bomb on top is even more dangerous as nails have been added to cause more potential injuries.

involving an explosion, the safe course is to suspect a terrorist attack.

SAFETY If one explosion has occurred, always assume that a second device is awaiting your arrival.

Any suspicious package should be treated as if it contains explosives, and it should be dealt with by a bomb technician. Never assume you can handle the package or remove it from the area. Handling bombs or suspicious packages is a very specialized field and should be left to people trained in that field.

The presence of chemicals or lab equipment in an unusual location such as a home or apartment is an indication of possible illegal activity. Three possibilities are: drug making, bomb making, or terrorism agent production, either chemical or biological. The most likely, based on statistics, is drug manufacture followed by bomb making. Although very common in some parts of the country, predominately the West Coast and the Pacific Northwest, illicit drug labs are not distributed evenly throughout the United States. Currently the number of drug labs, especially those set up to manufacture

methamphetamine, is increasing in the Midwest with law enforcement officers taking down more than twenty a month in some areas. This trend has made a progression from the West and is slowly moving to the East Coast.

In terms of terrorist activity, responders are most likely to encounter suspects making explosive devices. It is possible but highly unlikely that responders would locate a facility where people are attempting to make a terrorism agent such as sarin. The exception to the rarity of biological agents would be for the production of ricin, as the items required to make ricin are easily obtained and the production process is just as easy. Fortunately ricin does not have the easy potential to kill large numbers of people; it presents primarily an injection hazard, even though it is still very toxic through inhalation or ingestion. A responder in a small community or a rural setting is the most likely to run across any of these types of production areas. Several people are arrested each year for the possession of ricin, with the intent to use it for some type of criminal act. Most of the arrests are occurring in small towns throughout the United States. No matter what type of chemical threat responders locate, they should immediately isolate the area and call for assistance. The call should go simultaneously to the police, fire department, the HAZMAT team, and the bomb squad. This request would apply to all three types of labs—drug, bomb, or warfare agent—as any of these labs should be handled by both groups as a cooperative effort.

Another indicator of potential terrorism is an intentional release of chemicals into a building or the environment. Finding a chlorine cylinder in a courthouse would be unusual and should put responders on alert to the fact that there is a high probability that a terrorist may be at work. Finding chemical containers such as bottles, bags, or cylinders in unusual locations would also be suspect. In an industrial facility in which chemicals may be common, responders may discover an intentional release of a chemical.

NOTE One of the best indicators of a chemical release by terrorists will be a pattern of unexplained illness or injury.

A response to the mall for a seizure patient is not unusual. However, a response to a mall for six people having seizures is very unlikely and is more likely to involve a chemical release that could be from a terrorist attack. Victims twitching or having seizures, tightness in the chest, pinpoint pupils, runny nose, nausea, and vomiting are all signs and symptoms of a warfare agent attack. EMS providers will probably be the first group to identify the use of a chemical agent. Imagine arriving at an explosion, with large amounts of debris and twenty patients. To most, the injuries would appear as blast injuries, as they will in most cases be visible during a quick survey of patients. Other signs and symptoms, primarily the pinpoint pupils, probably will not be identified until patients are given a more thorough exam. Unconscious or dead patients with no obvious outward signs of death, such as trauma, may have been the victims of a chemical attack. This assumption must obviously be put in perspective. Responding to an apartment building in the winter in a northeastern city in which there are twelve unconscious people points in all probability to carbon monoxide poisoning. However, a call about twelve unconscious people in the mall during the summer would more likely indicate some type of chemical release. (When terrorists distribute a warfare agent via an explosive device to the building or surrounding area, it may not show signs of damage. In some cases the larger the device the more likely the detonation will consume the agent.) As a result most of the victims will not display signs or symptoms of a blast, although those closest to the device may suffer some of those types of injuries.

As with other responses, unusual odors or the presence of a vapor cloud may be an indicator that chemicals may have been released. Finding dead birds or other animals should put responders on alert to the chemical hazard and the potential for terrorism. It would also be unusual to respond to an office building or a house and notice unusual security measures of the type that you would not expect in that occupancy. Items such as extra locks, bars on windows, surveillance cameras, fortified doors, guards, and other unusual protection devices may provide a clue to possible illegal activity.

INCIDENT ACTIONS

A terrorist incident combines four types of emergency response into a large incident. These types of incidents have these four characteristics until proven otherwise.

NOTE Mass casualty EMS incident + Hazardous materials release and/or explosive devices + Crime scene considerations = Incident management challenges

---

The first responders can easily be overwhelmed and should commit to basic actions, such as life preservation. The IC will have enormous responsibilities dealing with all of the required actions. All of these components are difficult to handle individually, and now they are combined. The handling of a 100-person **mass casualty** incident is extremely difficult. It has the potential to overload the IMS system, and still it may only be one-fourth of a terrorist incident. Examine your response system, and determine how many patients present a concern. Some systems define a mass casualty as an incident involving five victims, while in other systems it may be ten to twelve. Obviously for a system with one EMS unit, more than one patient begins to cause system problems and may involve delays in treatment and transportation.

Imagine responding to an explosion at a mall, with 100 people injured. Such a **mass casualty incident (MCI)** will have EMS playing a predominate role. The police department, on the other hand, will be concerned with evidence preservation and crime scene considerations. Responders deal with the combination of these two priorities daily on a small scale. While responding to a crime scene, responders must take some extra time and consideration when treating patients. The possibility exists that the perpetrator(s) also used a chemical or biological weapon, and the explosion was the means of distribution. In such an incident, responders now are dealing with an MCI, crime scene, and potential chemical release with some, if not most, of the patients contaminated. On top of this scenario, add the potential for a secondary explosive device, one aimed at the responders—*us!* Each responding agency has major responsibilities. The HAZMAT team eliminates the chemical, while the bomb squad searches for a secondary device. EMS treats the patients, and the police department handles the criminal element.

The other aspect of a suspected terrorism incident is the massive response from the federal government, even if not requested. Later in this chapter more information is provided about available federal resources. In most cases the minimum response to a suspected terrorism incident would come from the FBI, initially from the local agent. There are 56 field offices across the country; almost every major city has a field office. Each FBI office has an agent assigned as the WMD coordinator who works with local responders. The FBI agent(s) typically arrive one to two hours into the incident. If terrorism is suspected, the FBI becomes the lead agency of the incident as is provided by a Presidential Decision Directive known as **PDD 39.** As with hazardous materials incidents it is important to know all involved players before terrorist attack. Knowing your local FBI agent can be very important, as meeting for the first time in front of an incident is not conducive to effective scene management.

Working out "who's in charge" concerns prior to an incident is crucial. This is important to figure this out with your local police, EMS, and emergency management agency prior to the incident. These and many other responders will be involved and have a variety of responsibilities at an incident, such as that shown in **Figure 7-7.** In general a unified command is recommended, although this does not mean command by consensus. All involved agencies supply input. The agency with the majority of the tasks to be accomplished is generally the group in charge. In the initial stages the fire/EMS authority would be in charge while rescuing victims, but after the victims are removed and evidence recovery is the next priority, then command may switch to law enforcement.

The response to assist with the incident can in many cases be the most difficult part of the incident to manage. The IC will be overtaken by a large number of federal agency representatives, who on a regular basis will be replaced by a later arriving supervisor. Response groups consisting of two to seventy responders may arrive uninvited, all trying to assist with the incident. The one group that will also overwhelm the system will be the media. In most cases the local media will react as normal and, for the most part hopefully, cooperate.

NOTE In any incident to which federal resources respond, rest assured that the national media will be following close behind.

Figure 7-7 Many local, state, and federal agencies will be involved in responding to a terrorist incident, each trying to accomplish similar goals. The use of an incident management system is essential to a successful outcome.

The national media do not know or follow local protocol and will require information. A lack of accurate information can be disastrous to an incident, as the media may use inaccurate information, which can cause deterioration of the incident and create more hysteria than may be present already. Media relations is critical in these type of events.

A crucial task at an incident perpetrated by terrorists is evidence preservation. As much care as possible should be taken to preserve evidence and make sure it is taken care of appropriately. The collection of evidence is primarily one of law enforcement's responsibility.

CAUTION Unless properly trained, fire service personnel should not collect evidence but should alert the police of its presence.

A whole host of issues goes along with the collection of evidence, including the chain of custody. This chain of custody, or the paper trail, that follows any evidence is crucial to the successful prosecution of the persons responsible. The failure to follow proper procedure or document the travel of evidence can result in a case being dismissed, regardless of any other evidence. In rare cases, such as the Murrah bombing, fire service personnel have been called on to collect evidence. If your agency is asked to collect evidence, it is best to assign a fire investigator or a fire marshal as they typically are trained in evidence collection. Another alternative is to double up and use a firefighter and a police officer to collect evidence.

It will take a cooperative effort to combat a terrorist attack. The primary function areas are rescue/life safety by fire and EMS personnel, hazard identification by the HAZMAT team, identification of possible secondary devices by a bomb technician, and incident management. It is important to communicate hazards to all personnel and to limit the response to essential personnel. Instead of having the whole alarm assignment report to the front of the building, use one or two companies to investigate while staging the other companies, away from dumpsters, mailboxes, or dead-end streets. When dealing with victims it is essential to isolate them until the cause is identified. The victims can have a large amount of information and should be questioned quickly. Questions to ask include:

■ What did you see, hear, or smell?
■ Is there one area that this was coming from or was it throughout the building?
■ Did you see anything else that may be suspicious?

■ What type of signs and symptoms do you have?

In addition, the police will need to conduct interviews. Documentation and preservation of the evidence will be essential in the successful prosecution of the terrorist.

GENERAL GROUPINGS OF WARFARE AGENTS

Terrorists can use a number of possible destructive warfare agents. These agents can be classified in three broad areas. The first is weapons of mass destruction (WMD), a term commonly used by the military. Some of the regulations that prohibit making, storing, or using terrorism agents are called WMD laws or regulations. Any item that has the potential for causing significant harm or damage to a community or a large group of people is considered a WMD. The other two classifications are nuclear, biological, and chemical (NBC) or biological, nuclear, incendiary, chemical, and explosive (BNICE). Both of these are descriptions of the types of materials that could be used in a terrorism attack. Although they have some slight variations, they all are used to describe the various types of agents that a terrorist could use. Most of the language differences come from funding legislation or a specific federal agency.

The military has devised a naming system for many of these agents, many of which are provided in **Table 7-1.** Responders should become familiar with these names, which are used in much of the literature and help guides. A military detection device will usually refer to the military name. When dealing with terrorism, we are entering another world that has its own language. The fire service must learn to adopt this new language. The broad areas of terrorism agents are further subdivided into several categories.

Nerve Agents

Nerve agents include tabun, sarin, soman, and V agent, all related to organophosphorus pesticides. They were designed for one purpose—to kill people. Although very toxic, the capacity of these agents to kill large numbers of people requires that the dissemination device function correctly, and a number of other critical factors must be in place for the attack to be effective. Although several gallons of sarin were used in the Tokyo subway attack in 1995, there was no effective method of distribution, and out of the thirteen people who died, it has been reported that the two people killed by sarin itself actually touched the liquid. The chemical and physical properties of these agents hinder their capacity to be effective as a stand-alone killer. To produce the toxic effect the agent must touch people in liquid form or be breathed while the material is in aerosol form. The material will not stay in aerosol form for very long, and it has a very low vapor pressure so it will not create vapors as standing liquid. All of the military nerve and blister warfare agents have a vapor pressure less than water, which means they do not evaporate quickly. Unless the liquid is touched or placed on the skin, it does not present a large hazard.

TABLE 7-1

Military Designations for Agents			
Name	Military designation	Name	Military designation
Tabun	GA	Sarin	GB
Soman	GD	Thickened soman	TGD
V agent	VX	Mustard	H
Distilled mustard	HD	Nitrogen mustard	HN
Lewisite	L	Hydrogen cyanide	AC
Cyanogen chloride	CK	Chlorine	CL
Phosgene	CG	Tear gas	CS
Mace	CN	Pepper spray	OC

SIGNS AND SYMPTOMS OF NERVE AGENTS

All nerve agents present the same types of signs and symptoms as organophosphorus pesticides, and in reality the difference is minor. Nerve agents are pesticides for humans and are a stronger, more concentrated version of commercially available pesticides. The signs and symptoms can be generally described using the term SLUDGEM, which stands for:

Salivation—victims will be drooling.
Lacrimation—their eyes will be tearing.
Urination—victims may lose control and urinate on themselves.
Defecation—victims may lose control of their bowels (diarrhea).
Gastrointestinal—they may experience nausea and vomiting.

Emesis—vomiting is another symptom.
Miosis—pinpoint pupils are also signs.

SLUDGEM describes all of the symptoms of exposure to a nerve agent, from the minor ones to the extreme signs. A slight exposure to any of the nerve agents will cause pinpoint pupils, runny noses, and difficulty breathing. A person who has come in contact with the liquid will be experiencing all the SLUDGEM signs in addition to convulsions. A person who is in convulsions needs immediate decontamination and medical treatment to survive. This treatment sequence must occur in less than ten minutes, and sufficient medication must be available on scene to accomplish the treatment. Most paramedic units carry enough medication to treat one or two patients who have severe symptoms.

Incendiary Agents

For the sake of classification we will place **incendiary agents** into the chemical classification, as chemicals are used in these devices. The most commonly used chemicals are flammable and combustible liquids. The standard Molotov cocktail is an example of an incendiary device that could be used by a terrorist. In some cases arsonists have used a mixture of chemicals, usually oxidizers to create very fast, high-temperature fires that could be used by a terrorist.

Blister Agents (Vesicants)

Vesicants, or more commonly called **blister agents,** include the chemical compounds mustard, distilled mustard, nitrogen mustard, and lewisite. These materials were never designed to kill, but were meant to incapacitate the enemy so that if one person was affected by one of these agents, several more would be needed to take care of the victim. The effects were to cause irritation and blistering to a severe degree. Although at high concentrations these materials can be toxic, their biggest threat is from skin contact and causing blistering. Their chemical and physical properties make them less of a hazard than the nerve agents. One large concern with these agents is the fact that the effects from an exposure are delayed sometimes fifteen minutes to several hours. Quick identification of a blister agent by the HAZMAT team is key to keeping the victims safe.

SIGNS AND SYMPTOMS OF BLISTER AGENTS EXPOSURE

One of the biggest risks with the blister agents is the fact that they may present delayed effects. If not detected early, patients might be released only to have problems later. In general blister agents are not designed to kill; they were designed to incapacitate the enemy, resulting in troops being assigned to assist with the wounded. It is possible to create scenarios in which fatalities could occur but these would be unusual cases. The effects from blister agents include irritation of the eyes, burning skin, and diffi-culty in breathing. The more severe exposure will result in blisters, which may be delayed. The only real street treatment for these signs and symptoms is decontamination and supportive measures. It is important that anyone who has had contact with the liquid blot the liquid off the skin and avoid spreading the agent. Luckily the chemical and physical properties of these agents make them difficult to disseminate, and coming in contact with the liquid would be the primary means of injury.

SIGNS AND SYMPTOMS OF BLOOD AND CHOKING AGENTS

Many of the blood agents are commonly found in industry and may be found at typical chemical facilities. The signs and symptoms of blood agents for a slight exposure include dizziness, difficulty breathing, nausea, and general weakness. With cyanides the breathing initially will be rapid and deep, followed by respiratory depression, usually leading to death. The two most common choking agents are widespread in industrial use, and chlorine can be found in every community. Signs and symptoms include difficulty breathing, respiratory distress, eye irritation, and, in higher amounts, skin irritation. Phosgene may present delayed effects, while chlorine's effects are immediate.

Blood and Choking Agents

In addition to being terrorism agents, the substances in this group are also common industrial chemicals. The first category, **blood agents,** includes hydrogen cyanide and cyanogen chloride. Both of these materials are gases and disrupt the body's ability to use the oxygen within the blood stream; they are referred to as chemical asphyxiates. **Choking agents** are even more common; chlorine and phosgene, are gases commonly used in industry. Chlorine is in almost every town, as it is used for water treatment and swimming pools. Any community that has a water system or swimming pool has some form of chlorine. Chlorine comes in containers ranging from 150-pound cylinders to 90-ton railcars; it also comes in tablet form (HTH™) typically used in residential pools. The release from a 90-ton railcar can result in hundreds of thousands of injuries and possibly many deaths as well, especially in an urban area. A small amount of chlorine can be very deadly or, at a minimum, create substantial panic in a community.

Irritants (Riot Control)

The most commonly used materials that are classified as potential terrorism agents are irritants, including Mace, some other form of pepper spray, or tear gas. The response to one of these types of incidents is common and usually impacts a large number of people. The usual target is a school, mall, or other large place of assembly. The use of one small container can immediately affect large numbers of people and make them immediately symptomatic. Luckily these materials are not extremely toxic, although they are extremely irritating, and the symptoms will usually disappear after fifteen to twenty minutes of fresh air. The response to one of these incidents is difficult as responders must treat patients with real medical problems and an unknown source that is difficult to identify.

Biological Agents and Toxins

The most likely agents that could be used in a terrorism scenario are **biological agents** and **toxins.** Some of the materials in this grouping include anthrax, mycotoxins, plague, tularemia, and ricin. Out of all of the agents for a terrorist to make, this grouping is the easiest, especially ricin. The fatal route of entry differs with each of the agents and may be skin contact, inhalation, or injection. These agents are difficult to distribute effectively, and in some cases exposure to sunlight may neutralize many of these agents. The two most commonly used agents in this category are anthrax and ricin.

Anthrax is a naturally occurring bacterial disease that is commonly found on dead sheep. This naturally occurring anthrax is contagious through skin contact or by inhalation of the anthrax spores. Although relatively easy to obtain, it is more diffi-

---

SIGNS AND SYMPTOMS OF IRRITANTS

The signs and symptoms for a slight exposure to a high dose are the same with the exception of increasing severity. The signs and symptoms are eye and respiratory irritation. There is no real treatment except removal to fresh air; in fifteen to twenty minutes the symptoms will begin to disappear. Supportive care can be provided.

cult to culture and grow the proper grade of anthrax. To produce fatal effects the type of anthrax required is called military grade anthrax, which is very difficult to produce. Even if developed, it must be distributed effectively and under the right conditions.

Although ricin is easy to make, the required distribution method makes a large-scale incident unlikely as it must be injected to be truly effective. Ricin is 10,000 times more toxic than sarin; a small amount of 1 milligram, about the size of a pin head, can be fatal. Death will usually occur in several days after the injection. By other routes of entry such as inhalation or ingestion, the most likely consequence is that many people would get sick but would eventually recover. Next to explosives, ricin is the leading choice of domestic terrorists; several times a year someone is arrested for possession of ricin.

The most successful attack using a biological material involved the use of salmonella bacteria by the Rajneesh cult in Oregon. In an attempt to influence a local election, members of the cult put the material in the salad dressings at several fast-food restaurants. More than 715 people fell ill from the deliberate contamination.

Radioactives

Nuclear agents, unfortunately, must be included on the list of possibilities that a terrorist could use. There are two type of radiation events, **nuclear detonation** and **radiological dispersion device (RDD).** The use of an actual nuclear detonation device is unlikely and very improbable given the security these materials have. The amount required and the specific type also makes use unlikely. There is current unsubstantiated speculation that some small nuclear devices may be missing from Russia. A more likely threat is that some nuclear material, perhaps originating from Russia, could be used in an RDD. An RDD is a device that disperses radiological materials, usually through an explosive device. Using a pharmaceutical grade radioactive material attached to a pipe bomb could cause a large amount of radioactive material to be distributed. The perception to the public and to

RESPONSE TO ANTHRAX HOAXES

With the number of anthrax hoaxes on the rise, emergency responders should establish a plan for response. From the outset it would be difficult for someone to manufacture and successfully pull off an anthrax threat. In many of the hoax incidents people received letters proclaiming that they had been exposed to anthrax. In some of these cases the only thing present in the envelope was a letter. Anthrax is not invisible, and a quantity of the material must be present if the attack is to succeed. To harm victims, the material must be inhaled, and to be inhaled, it must be distributed into the air. Just the placement of the material in an envelope does not present a high risk to the letter opener. To pose a risk, the solid material must be dispersed into the air, and if no dispersal occurs, the risk to people in a building is very small. The only action required is to double- or triple-bag the envelope, package, or material in bags suitable for evidence collection. Procedures for the collection of evidence should be followed, and the local police and FBI should be consulted.

People who touched the material should be instructed to wash their hands with soap and water. They do not require full body decontamination, nor are special solutions required such as a bleach and water mixture. The military recommends that humans exposed to anthrax rinse thoroughly with water. Anyone who was within several feet of the person opening the package does not require decontamination. The only reason for a full-body wash would be if victims spilled the material on themselves from head to toe, but no immediate action is required. Victims can be taken to a shower and be provided privacy as they wash. People directly exposed to material that may contain anthrax should be entered into the health care system so that they can be advised of the signs and symptoms of exposure and have emergency contact information relayed to them.

There is no need to start prophylactic antibiotics on people just because they opened a letter with a powder in it. There is sufficient time to complete lab analysis on the material before medical treatment is required. The threat from a solid material (no matter how toxic) is from inhalation of the dust or touching the material. Gloves and SCBA are adequate protection for collection of the evidence. The FBI has a number of labs around the country set up to assist with the identification of WMD agents, and the local FBI office should be contacted for assistance. The FBI has established a National Domestic Preparedness Office (NDPO) to assist first responders with planning and training for a terrorist attack.

many responders is that this would be a radioactive disaster. Radiation meters would indicate that there was radioactive materials spread around. The reality is that the amount of radiation would not be dangerous, and as time passed the danger would lessen, as the radioactive material becomes less hazardous. Unless put in perspective with good public information, an RDD threat would cause tremendous concern in the community. Radiation is a big unknown and causes fear because it is unknown. That makes it a prime weapon for a terrorist.

Other Terrorism Agents

While considerable emphasis is placed on warfare agents, in reality many other common industrial or household materials can be just as deadly, if not more so.

NOTE Remember that a terrorist wants to create panic, not necessarily kill or injure massive numbers of people.

The most likely scenario is a pipe bomb, followed by the use of ricin. Other industrial chemicals can also be used, many of which are just as deadly, if not more so than most warfare agents. Responders should not get lulled into a false sense of security if they do not find a warfare agent, or if the device is only a small pipe bomb and not a moving truck filled full of ammonium nitrate and fuel oil. Nerve agents anthrax, or other chemical warfare agent have only been used twice each in Japan and have never been used in the United States, nor manufactured by anyone other than the military. The FBI Bomb Data Center reports that out of the 3,000 bombings that occur each year there have only two large bombings; however, these "small" bombs kill an average of 32 and injure 277 people a year. Although responders should be on the watch for secondary devices, an eleven year history of bombings shows only 4-5 secondary devices out of an approximate 33,000 bomb incidents.

DETECTION OF TERRORISM AGENTS

The detection of terrorism agents is difficult and, given the potential circumstances, must often be done under severe conditions. The response to terrorism incidents has changed how many HAZMAT teams operate and has increased their capability to handle other situations. The confirmation that a terrorist agent has been used is difficult; as a group they are extremely hard to detect. The exceptions are standard industrial materials, such as chlorine, which are easy to detect and confirm. The only current methods of detecting warfare agents are adapted from the military. Unfortunately these detection devices were made for a military environment and not civilian situations. The devices were built to detect warfare agents and designed not to provide false positives for chemicals commonly found on the battlefield. They could, however, provide false positives for some substances common in other environments. Even though these devices are being used in the civilian world, they have not been reconfigured to eliminate these false positives. Many substances that emergency responders see on a regular basis, such as paint thinner, would be unusual in a field of battle and might cause a military detection device to alert for nerve agent.

Electronic detection devices, dip stick type tests, and wet chemistry tests have been designed to detect these agents. One of the more difficult agents for which to provide street level detection is the biological group; given current technology, it is recommended that the sample be sent to an FBI-designated lab for analysis. In some cases standard civilian detection devices such as a photoionization detector will also play a role in the detection of terrorism agents. Other tests used by HAZMAT teams have applicability in the detection of terrorism agents. Due to the many mitigating factors the detection of these agents will be difficult and may require lab tests to be conclusive. For more information on detection devices for terrorism see Delmars text *Air Monitoring and Detection Devices*.

FEDERAL ASSISTANCE

The federal government has established roles and responsibilities in case of a terrorist event, and these roles are provided in Presidential Decision Directive 39 (PDD 39). This document names the FBI as the lead agency during the emergency (crisis management) stage of the incident. The Federal Emergency Management Agency (FEMA) becomes the lead when the incident is no longer in the emergency phase (consequence management). The FBI has a Hazardous Materials Response Unit (HMRU), which on a potential terrorist incident provides identification, mitigation assistance, and evidence collection. The HMRU is a multifaceted group that responds not only to terrorism incidents, but also to scenes involving explosives, drug labs/incidents, and environmental crimes. Since its inception, the HMRU has been responding to an ever-increasing amount of incidents. The majority of these incidents involve biological threats, although a number of incidents involve other materials. HMRU is also pre-positioned at a number of events throughout the nation and works closely with local HAZMAT teams.

A response from FEMA will vary with the incident, and the number of FEMA personnel will resemble a response similar to any other disaster to assist in restoration and recovery issues. The Urban Search and Rescue (USAR) teams fall under FEMA and are activated by following the emergency management chain from local to state, to FEMA. At this time there are 25 USAR teams across the country. They provide expertise in heavy rescue operations such as building collapses. They are a multifaceted team of 62 people and are comprised of rescue specialists, dog search teams, medical specialists, communications specialists, and an engineering and rigging component. The greatest majority of the victims are rescued by the first responders and other local specialized resources. But in some cases trapped victims may be alive for many days and may require the expertise and equipment of a USAR team. One unfortunate aspect is that the USAR teams have a delayed response. It will take them several hours to become airborne, not including travel time. If there is a possibility they may be needed, their assistance should be requested early. Like many teams, USAR may send an advance party hours before the arrival of the remainder of the team.

A number of other agencies may be involved in the response to a terrorist incident. The military has a couple units with terrorism response capabilities and responsibilities. The Army has the Technical Escort Unit (TEU), which comes from the Aberdeen Proving Grounds just north of Baltimore. The TEU is assigned to provide escort service for warfare agents and be the troubleshooting group in the event of an incident involving military warfare agents or explosives. When an incident involves potential chemical agents, the TEU will respond at the request of the FBI to assist with identification and mitigation of the incident. The TEU handles chemical, biological, and explosive materials and other hazardous materials. This self-contained unit employs the lab resources of the Soldiers Biological and Chemical Command[2] (SBCCOM).

The Marines have a unit known as the Chemical and Biological Incident Response Force (CBIRF), based in Camp Lejeune in North Carolina. This unit has responsibility for the response to terrorism across the world. Its three main components are decontamination, medical, and security. The CBIRF is a self-contained unit and requires only a water source for long-term operations. It can also provide detection and mitigation. CBIRF responds nationwide upon request to terrorist incidents. All of the federal resources, including CBIRF and TEU, can and have been pre-positioned for certain events, such as the Olympics, political conventions, and visits by dignitaries. These resources integrate within the local system and are immediately available, usually hidden away from the public.

Other federal programs include training under the Nunn-Lugar-Domenici legislation, which was passed in September 1996 (PL 104-201). The law, known as the Domestic Preparedness Training Initiative, mandated that the Department of Defense provide training to the 120 major cities across the United States. It was intended to enhance the capability of the local, state, and federal response to incidents involving NBC materials. Part of this process is an assessment after the training to determine the cities' capabilities. A second component of the law provides for twelve teams to be fielded by the National Guard in the FEMA regions across the United States. These teams, known as Rapid Assessment and Initial Detection teams (RAID), are being established to provide a National Guard component to the response to terrorism.

BASIC INCIDENT PRIORITIES

Rescue live victims using full protective clothing and SCBA.

■ All terrorist agents are predominantly hazardous through inhalation. Avoid touching any unknown liquids or solids, as most are also toxic through skin contact.

■ If there are live victims in an area, then the atmosphere cannot be that toxic.

Use a quick in–quick out approach.

■ Do not treat victims; remove them from the area.
   ■ Swoop and scoop is the order of the day.
   ■ Any possible fatalities should be left in place, as they are considered evidence.
   ■ Remove the live victims and then check the vital signs outside the hazard area. Do not take the time to treat the victims in the hazard area.
■ Keep in mind the terrorist may be among the injured.
   ■ Ensure police presence in and around the event.
■ Watch for secondary devices.
   ■ Use bomb-sniffing dogs to search for secondary devices.
   ■ Avoid staging all equipment together.

Request HAZMAT and bomb squad assistance. The sooner you can eliminate the potential for chemical agents or a secondary device, the better off you will be. The HAZMAT team may have the ability to detect terrorism agents, and most HAZMAT teams are working with bomb squads on a more frequent basis.

Limit personnel operating in the hazard area.

Establish multiple staging areas, out of the line of sight.

Notify your emergency management agency so that it can mobilize the state and federal resources.

If a building has collapsed or there is potential for a building collapse, request a tactical rescue team or a USAR team to assist.

Isolate all victims, separating contaminated from clean victims.

Establish a safe triage, treatment, and transport area—away from the impact or hot zone.

Notify all area hospitals of the incident.

Remember that the incident is a crime scene and make provisions to preserve as much evidence as possible.

If you suspect chemical agents, use the DOT NAERG or other reference sources such as the *Management of Chemical Warfare Agent Casualties* guidebook to guide you through safety precautions and patient treatments.

SUMMARY

The response to a potential terrorism incident can be very challenging, and every responder must be aware of its possibility. To be safe, wear all PPE. Do not delay in the environment, and once the live victims are out, relocate to a safe area. Be aware of the potential for secondary devices, and request HAZMAT and bomb squad assistance quickly. There is potential for one to thousands of patients who may be injured or dead. Some scenarios involve tremendous loss of life, including a large number of responders. There exists the possibility that we may lose and the terrorist may win, but we can avoid this outcome by training, planning, and preparing for such an incident.

KEY TERMS

Anthrax   Naturally occurring biological material that is severely toxic to humans; common agent cited in hoax incidents.

Biological agents   Toxic substances that are living materials or are obtained from living organisms.

Blister agents   See *vesicants*.

Blood agents   Chemicals that affect the body's ability to use oxygen and in many cases prevent the body from using oxygen, producing fatal results.

Butyric acid   Fairly common lab acid used in many attacks on abortion clinics; although not extremely hazardous, it has a characteristic stench that permeates the entire area where it is spilled.

Choking agents   Chemicals that produce an immediate reaction (coughing, difficulty breathing); terrorism agents that are considered choking agents are chlorine and phosgene, both very toxic gases.

Incendiary agents   Chemicals used to start fires, the most common being a Molotov cocktail.

Mass casualty   Incident in which the number of patients exceeds the capability of the EMS system to manage the incident effectively. In some jurisdictions this can be two patients, while in others ten may qualify as a mass casualty.

Nerve agents   Very toxic materials used by the military to attack opposing troops; can be equated to high strength pesticides.

Nuclear detonation   Explosion caused by a nuclear device.

PDD 39   Presidential Decision Directive 39, which established the FBI as the lead agency in terrorism incidents; the FBI is responsible for crisis management; and FEMA is the lead for consequence management.

Radiological dispersion device (RDD)   Radioactive material that is coupled with an explosive device, which spreads the radioactive material. It is not a nuclear detonation, as the explosion is caused by a standard explosive charge.

Ricin   Biological toxin that can be used by a terrorist or other person attempting to kill or injure someone; the easiest terrorist agent to produce and one of the most common.

Toxins   Substances that are biological in nature and are obtained from naturally occurring materials.

Vesicants   Group of chemical agents that cause blistering and irritation of the skin; commonly referred to as *blister agents*.

Weapons of mass destruction (WMD)   One of the
terms used to describe military, chemical, or biological weapons systems.

REVIEW QUESTIONS

1. What are four potential targets?
2. Name four indicators of potential terrorist activity.
3. What are the most common agents that are readily available and could be used in a terrorist attack?
4. Who are the three main local/regional agencies or groups that should be notified immediately of a suspected terrorist attack?
5. What is the process for requesting federal assistance?
6. Out of the BNICE agents, which is the most likely to be found at an incident?
7. What is the second most likely agent?
8. Which agent is designed to kill immediately?
9. What are the immediate signs and symptoms of sarin exposure?

ENDNOTES

[1]Roe vs. Wade was the U.S. Supreme Court decision in 1973 that allowed legalized abortions. This case has created much of the turmoil on both sides of the abortion battle.

[2]Formerly known as the Chemical and Biological Defense Command (CBDCOM).

Suggested Readings

Bevelacqua, Armando, and Richard Stilp, *Terrorism Handbook for Operational Responders,* Delmar, a division of Thomson Learning, Albany, NY, 1998.

Buck, George, *Preparing for Terrorism: An Emergency Services Guide,* Delmar, a division of Thomson Learning, Albany, NY, 1998.

Hawley, Chris, *Air Monitoring and Detection Devices,* Delmar, a Division of Thompson Learning, Albany, NY 2001

Smeby, L. Charles (editor), *Hazardous Materials Response Handbook,* 3rd ed., Supplement 14, *Emergency Response to Incidents Involving Chemical and Biological Warfare Agents,* Lt. Col. John Medici (ret.) and Steve Patrick, National Fire Protection Association, 1997.

CHAPTER 8

HAZMAT AND LAW ENFORCEMENT

OUTLINE

STREET STORY

Two stories are provided with regard to law enforcement officers and hazardous materials, both the result of tunnel vision on the part of the officers. The first involves a direct chemical exposure in which two officers who were staking out a farmer's field were sprayed by a pesticide. Fortunately, after being sprayed the officers had the foresight to seek medical attention. As a general rule pesticides are designed to kill, although their toxicity to humans varies from compound to compound. These two officers were becoming symptomatic from their exposure and were having a reaction. One of the common reactions is difficulty breathing, followed by the cessation of breathing. Luckily the medical crew recognized the signs and symptoms and took appropriate action. The actions required to save the officers' lives were to strip them, rinse them off, and scrub them for a period of time, then medicate them and send them to the hospital for further decontamination and treatment.

The other story relates to a traffic stop that was a couple of weeks after the bombing of the Alfred P. Murrah building in Oklahoma City. At about 3 A.M. a patrol car stopped a rental truck on a residential street just off a main road for erratic driving. The officer approached the vehicle and when he talked to the driver the officer noted the driver's slurred speech and that there was a chemical odor coming from the truck. Another officer arrived to assist and decided to open the back of the truck. In the back of the truck were about fifteen plastic 55-gallon drums, none of which were marked. Although a lot of details were not known at the time about the bomb used at the Murrah building, it was known that it was a rental truck with 55-gallon drums in the back of the truck. The officers removed the driver of the truck to a patrol car and called for the fire department. When the fire engine arrived they noticed the contents of the truck, quickly backed up and requested that a hazardous materials unit respond. Through radio communication the engine asked that the suspect remain on the scene and that the officers who had been in proximity to the truck remain isolated. The HAZMAT crews arrived and concurred with the engine company that the drums in the back of the truck were a scary combination and a little too close to the McVeigh bombing. Of utmost concern were the contents of the drums. The HAZMAT crews wanted to speak to the driver of the truck both to gain information and to determine if his erratic driving was caused by the fumes in the drums or from the suspected alcohol intake. The HAZMAT crews were unable to locate the driver of the truck as he had been taken to the local precinct, along with the first arriving officers. It was not known at this point if the suspect was contaminated, if he had cross-contaminated the arresting officers, or if the contamination had spread. Other HAZMAT crews had to be requested to respond to the precinct to assess that situation, while the first responding HAZMAT crews had the task of seeing what was in the truck. The contents of the drums were a mixture of waste pesticides and insecticides. The suspect would collect various chemicals and place them into a drum mixed with other chemicals. He was operating a pesticide removal business illegally and was not following any labeling, storage, or transportation regulations, creating a potentially dangerous situation. It was determined that his erratic driving was probably a combination of alcohol intake and chemical exposure, since he was over the DUI legal limit and did need medical treatment for his exposure. The officers involved were also decontaminated and treated at the hospital.

—Chris Hawley

AUTHOR'S NOTE Although this section is labeled for law enforcement officers, it has applicability for all emergency responders and should be included in all training programs. The material in this section would be considered at the operations-plus level.

OBJECTIVES

After completing this chapter, the reader should be able to identify and explain:

■ Types of HAZMAT situations that are common to law enforcement (A)

■ Why the use of universal precautions is important (O)

■ Types of labs that may be encountered (O)

■ Special requirements for SWAT operations (O+)

■ Considerations at suspected bomb incidents (O)

■ Considerations for evidence collection in hazardous environments (O+)

INTRODUCTION

Getting shot at is an immediate and very real threat to a law enforcement officer's life. A chemical exposure can have that same type of immediate threat, but more commonly it may present a long-term hazard that may take years to take a law enforcement officer's life. Just entering a suspect's home to make an arrest places the officer in danger of a chemical exposure that could have long-term effects. Law enforcement officers in the United States have always had some involvement with hazardous materials, however, it was usually limited to traffic control or security issues. In the West they encountered drug labs on a frequent basis, which are very dangerous operations. These labs are now common in the Midwest and are becoming more frequent in the East. Even without drug labs law enforcement officers are getting more involved with hazardous materials response. In some states the local or state law enforcement agency is the regional HAZMAT team. In other cases law enforcement agencies are training certain components of their departments for some type of HAZMAT response. What most law enforcement officers are going to learn from this chapter is the fact that they have been involved with hazardous materials for many years, but the question that needs to be asked is whether they were properly prepared and protected.

COMMON HAZMAT INCIDENTS FOR LAW ENFORCEMENT

When law enforcement officers respond to auto accidents, medical calls, fires, or known chemical emergencies they are at risk for chemical exposure. All of these present some type of risk for the law enforcement officer. In many cases the law enforcement officer is the first arriving responder. The actions provided in Chapter 5 should be followed by this first arriving responder.

Actions taken by the first arriving responders who may not have appropriate PPE should be limited to isolating the event, requesting appropriate resources, and then coordinating the arriving resources.

SAFETY Since many law enforcement officers have limited protective clothing they should avoid all types of chemical exposures.

SAFETY Going into a chemical situation without respiratory protection is like going into a gun battle without bullets or a bullet-proof vest.

All responders regardless if they are fire, EMS, or law enforcement officers require some form of HAZMAT training. Chapter 1 provides more detailed information on the training levels, but the specialty areas described in this chapter require some additional training. Just entering a building that has had or potentially could have a chemical release requires a minimum of operations-level training. If the officer is going to collect evidence that may involve hazardous materials then some additional training is required, which may involve the officer becoming a HAZMAT technician. The investigators who are handling the drum dump, which is an environmental crime, as shown in Figure 8-2, are required to have HAZMAT training. SWAT operators who may enter a chemical environment require operations training geared toward their work tasks. OSHA requires that all employees work safely and that job-specific training be provided to the employees. This training also requires annual recertification, which must be provided by the employer.

Figure 8-2 Investigators at this drum dump exercise are developing the basis for a sampling plan. They are using air monitors, Polaroid photographs, a site map, and container listings to document the site.

Blood-Borne Pathogens

Probably the most common exposure to hazardous materials for law enforcement officers is to blood-borne pathogens. The materials in this category include blood, saliva, urine, and other bodily fluids. Hopefully with the awareness of the dangers of hepatitis A, B, and C, human immunodeficiency virus (HIV), and tuberculosis law enforcement officers will take universal precautions for these materials. Hepatitis C is now becoming more common, and many emergency response workers have suffered some serious health effects from being exposed to hepatitis C. The issue of tuberculosis is one that is probably not thought of by law enforcement officers. Being in close proximity to a victim who is coughing, and is later determined to have tuberculosis, will place the officer in danger to contracting tuberculosis. Fortunately, the process of protecting oneself against these pathogens is relatively easy, just by using universal precautions.

NOTE In 1991 OSHA issued the Blood-Borne Pathogens Regulation, which requires that workers who may be exposed to these types of materials be trained in the hazards of these materials and be provided protective equipment.

Drugs and Chemical Effects

The next biggest exposure issue for law enforcement officers is most likely drug related. Many drug users use chemicals as part of the process to get high, and the drugs themselves are usually toxic, which may present other hazards. The next section provides information related to drug labs so we will not discuss them in this section. Drug users may use a combination of things to get themselves high, and many of these items are toxic and flammable. In some cases ether will be used to assist in the heating process of several types of drugs. This material in pure form is extremely flammable, and the container may eventually become a shock-sensitive explosive. Most drugs are in solid form, which means that they present little risk unless eaten or touched with bare hands. In some cases the drugs may be stored or used with flammable and toxic liquids.

The most common time that law enforcement officers could be directly affected by drug use is when a person is **huffing.** When someone is huffing they typically use a toxic and/or flammable material. Many common household items such as paints, glues, hair spray, and solvents provide the high that some people desire. The HAZMAT response to this action is that these materials are toxic and flammable and are now in the air. In most cases the user will spray the material into a bag to concentrate the vapors, but when they are done the vapors will remain in the room of use.

SAFETY Officers must be careful with the evidence from these crimes, as a patrol car is a confined space with limited air movement and the fumes could cause problems for the officers in the car.

CLANDESTINE LABS

There are several types of labs that emergency responders are likely to encounter. The most common would be the drug lab, but other possible labs include **explosives labs, chemical weapons labs,** or **biological weapons labs.** All may be booby-trapped or be set up

UNIVERSAL PRECAUTIONS

In many states EMS providers, law enforcement officers, and firefighters are all covered by this regulation. Part of the regulation required testing for some materials, exposure reporting programs, and the development of an infection control program. The use of **universal precautions** is to prevent the transmission of infectious material from the patient to the care provider. Being universal, emergency responders always wear some type of protective clothing, usually gloves, when dealing with all patients. A big part of your protection against these materials involves normal hygiene practices, and simply washing your hands helps prevent problems in the future. In situations where fluids are present added protection is necessary such as glasses or goggles and a gown or coverall may also be necessary. OSHA developed specific regulations for workers who may be involved with tuberculosis patients. When dealing with known or potential tuberculosis patients respiratory protection is crucial for the responders' protection.

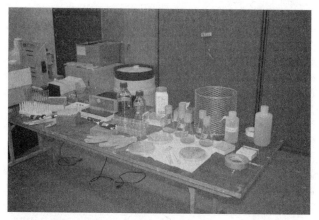

Figure 8-3 Much of the equipment at a biological agent lab will resemble the equipment shown here. There will be petri or culture dishes and flasks, and incubators. Some of the culture dishes may have a red gum like substance known as red agar to facilitate the culture growth. In some cases this lab will be isolated from sunlight, and will be temperature controlled.

to harm responders. A key safety note is that a booby trap does not know if the person coming through the door is a police officer, firefighter, or EMT. For the most part a biological weapons lab, which is shown in Figure 8-3, can run unattended without any major concern and may be shut down in any number of ways without major consequence. On the other hand, a drug, explosives, or chemical weapons lab should only be shut down by someone trained to do so, since these labs are especially dangerous.

SAFETY All labs have their inherent dangers to responders, and all are very dangerous locations to occupy.

Drug Labs

There are many types of drug labs. For example, there are more than six common methods of production for methamphetamines. In 1973 the DEA discovered forty-one labs, and in 1999 they discovered 2,155 labs. Of the lab seizures by the DEA almost all were meth labs. The map shown in Figure 8-4 provides proof that the labs are moving eastward at a rapid rate and soon all of the United States will have to deal with this problem. Due to its popularity meth production is becoming common, however, it does involve a dangerous production process. The production of drugs requires the use of many chemicals. These chemicals can be purchased outright, stolen, or manufactured using other chemicals. Because many drug-producing chemicals are hot listed or cannot be purchased the producer must resort to innovative methods to produce the chemicals. The "Nazi" method of meth production involves the use of anhydrous ammonia, which is usually stolen from a chemical facility. If a 150-pound cylinder of anhydrous ammonia developed a leak or catastrophically failed, people downwind for miles could be affected, and those in the immediate area would be in grave danger. Some of the common chemicals used in drug labs are found in Table 8-1.

Drug labs can be very complicated to set up and can be found in any number of locations such as homes, barns, hotels, storage units, and even trucks. Emergency responders routinely encounter these labs inadvertently through other responses. Responders who encounter a drug lab should notify their HAZMAT team and bomb squad and the local office of the DEA. The shutting down of a drug lab is a complicated and very dangerous process.

It is this heating and cooling process that indicates the type of lab that may be present. When

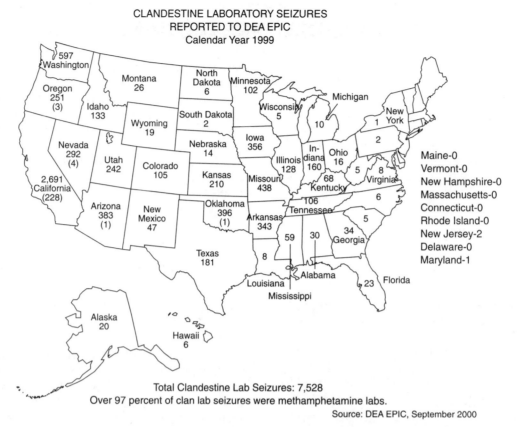

CLANDESTINE LABORATORY SEIZURES
REPORTED TO DEA EPIC
Calendar Year 1999

Total Clandestine Lab Seizures: 7,528
Over 97 percent of clan lab seizures were methamphetamine labs.

Source: DEA EPIC, September 2000

Figure 8-4 This map shows the locations of the labs that the DEA took down in 1999. The number in parenthesis is the number of super labs, which were producing large quantities of drugs. The map clearly shows the labs moving eastward into almost every state, with some states in the West having thousands of labs being taken down.

TABLE 8-1

Common Drug Lab Chemicals

Flammables—Ethers (diethyl ether, petroleum ether), ethanol, isopropyl alcohol, and acetone

Corrosives—Hydrochloric acid, methylamine, hydriodic acid, sulfuric acid, nitric acid, sodium hydroxide, piperidine, hydrogen chloride gas, and anhydrous ammonia

Other chemicals—Mercuric chloride, acetic anhydride, and cyanide

Chemicals used to produce methamphetamines (various methods)

Acetic acid	Hexane
Acetic anhydride	Hydriodic acid
Ammonium acetate	Hydrochloric acid
Anhydrous sodium acetate	Iron filings
Benzaldehyde	Lead diacetate
Butylamine	Mercuric chloride
Ephedrine hydrochloride	Nitroethane
Ethanol	Phenylacetic acid
Ethyl ether	Pyridine
Freon	Red phosphorus

COMMON CHEMICALS USED FOR EXPLOSIVES

Acetone	Graphite	Potassium nitrate
Aluminum powder	Magnesium	Potassium perchlorate
Ammonium nitrate	Nitric acid	Sodium chlorate
Ammonium perchlorate	Phenol	Sodium nitrate
Ethyl ether	Picric acid	Sulfuric acid
Glycerine	Potassium chlorate	Urea (or urea compounds)

responders see glassware that is distilling chemicals, in other words evaporating a certain component, and then rehydrating or condensing another portion of the original chemical this is indicative of possible drug production. The end result of the process results in a solid form, usually a powder. There will be a production line approach to the glassware, with some solutions being heated while others are cooled. At some point during the process gas cylinders may be used, which allows the gas to mix with some other part of the process. Many of the chemicals involved in the production of drugs are flammable, and although many are also toxic the predominant hazard is flammability.

For the materials that are toxic, as long as SCBA is worn and materials are not touched or eaten with bare hands they cannot present harm to the responder. It is highly recommended that personnel from the police and emergency services departments receive training regarding drug lab awareness, with some receiving specialized training in this area.

SAFETY When chemicals are being heated or cooled there is a chance for a violent reaction.

SAFETY When choosing protective clothing to enter a lab it is important to protect responders against the predominant hazard.

Explosives Labs

Although not common it is possible that law enforcement officers would encounter an explosives lab, which is the predominant weapon of choice for a terrorist. An explosives lab can run from a workbench pipe bomb builder to a full chemical production facility. A person building a pipe bomb does not need much equipment other than some simple hand tools, pipes, caps, and some powder. A person making RDX or other more technical explosives will need some more equipment. Depending on the availability of some materials, the bomb maker may have to make some chemical components as opposed to purchasing them. It is when materials are made at home that the danger increases for a responder. Many of the processes for making the chemical components for explosives are very dangerous and present a lot of risk for the builder and responders. An explosives lab differs from a drug lab in the fact that most of the processes do not involve heating, cooling, condensing, or distilling. However, some processes, such as those used to make some of the chemicals, do use some of these procedures, so there are no black-and-white rules for identification. The major work at a explosive lab is usually mixing of materials, in most cases solids and liquids. If gases are used they are usually coupled with the explosive device and may be used to increase the heat of the explosion, boosting its efficiency. Someone who is making explosives will usually have a large amount of powders in their house, and will have a large number of chemicals such as those listed in the box "Common Chemicals Used for Explosives."

Some of the other indicators of an explosives lab will be the presence of ignition devices, boosting charges, or blasting caps. Many people who play with explosives manufacture are doing so to make fireworks and will have cardboard tubes for the explosives to be placed into. In one case it was originally thought that the explosives maker was just making fireworks, but some other explosive devices were found to have BBs and nails taped and glued to the outside of the explosive. Neither of these have an effect on the display ability of the explosive device and are only designed to kill or maim. When an explosives lab is discovered the local bomb squad and the local HAZMAT team should be called in to assess the materials. Since the stability of many of the materials cannot be assured, it is usually necessary to destroy many of the found materials in a remote region.

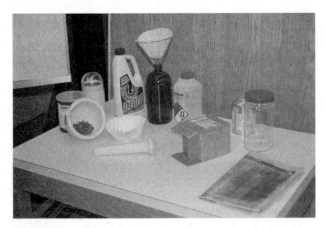

Figure 8-5 Shown above is an example of what a ricin lab may look like, with some representative equipment and chemicals.

Terrorism Agents Labs

These are the least likely encountered labs, but responders must be aware of their possibilities and of the unique features of these types of labs. The two types of terrorism agents or WMD labs are chemical weapons or biological weapons. Statistically the most likely lab is a biological toxin lab, which may be used to manufacture ricin, as shown in Figure 8-5. The FBI arrests about six people a year for the possession of ricin, usually accompanied by a threat. Biological labs may be set up to attempt to make other biological materials. Other than ricin, botox, and a few other biological materials the manufacture of biological weapons is very difficult. Chapter 7 provides a box "Assessing Terrorist Threats" about this subject. The manufacture of ricin is a simple process requiring only a few items, such as castor beans and some readily available chemicals. The process for making some of the more advanced biological agents such as anthrax is more difficult and requires a higher level of education, sophisticated equipment, and access to raw materials. The development of biological agents involves culturing of the material, usually in a petri dish. These petri or culturing dishes may be kept in an incubator or oven-like device to keep the material warm and at a constant temperature. Depending on the type of agent there may be grinders, dryers, and sieves present to finish off the product. The major route of entry for many of these products is through contact or ingestion. The only inhalation concern would be during the grinding process, but a simple HEPA mask offers more than enough protection for a responder. Some of the materials used as part of the process may be flammable with some

toxicity, but the chemicals used to clean glassware and tools are usually highly corrosive and in most cases will be sodium hydroxide (lye) and/or bleach. A bio weapons lab will involve some chemicals, but in some cases it resembles a greenhouse more than a chemical lab. Some biological materials are sensitive to light and will be kept in the dark, therefore, the characteristic distilling and condensing glass apparatus will be missing.

A chemical weapons lab will have two methods of production: the development of new chemical agents through standard production methods or the synthesis of existing materials. The development of new chemical agents is the more difficult of the two processes, and is nearly impossible except for someone with a chemistry background and access to some chemistry equipment. The recipes for chemical weapons are fairly sophisticated and require access to many raw materials that are hot listed. There have been only two arrests of persons in the United States for attempting to make a chemical weapon such as sarin, a nerve agent. Both were arrested for ordering the raw materials, and it is thought that the two individuals did not have the mental capacity to manufacture the agent. The production of these agents can be very risky to the people involved, and without safety precautions they may end up killing themselves in the process. Some of the off gases from the production of warfare agents are extremely dangerous.

The most probable scenario for the development of a chemical warfare agent is through synthesis of an existing product. The criminal would take existing materials, which are usually in diluted form, and synthesize them or reduce them down to a concentrated product. There are many organophosphate pesticides on the market that are applicable to this type of scenario. The standard household pesticide usually has 0.5 or 0.05 percent of pure product mixed with an inert ingredient. A mixture that is used by a farmer may be in the order of 40 to 50 percent pure form, which is then diluted in the farmer's tank. In order to be more effective on humans the pesticide must be concentrated and not diluted. The criminal must devise a method of removing one or more of the inert ingredients. Since the inert ingredient is usually flammable or combustible, this is not that difficult of a task. A mechanism must be devised to off-gas the inert ingredient and then capture the pesticide and collect it.

Pesticides that are already concentrated are known as technical grade pesticides. These technical grade pesticides are in pure form and are not diluted in any way. These materials are not in common use, but can be found in and around the

United States. One type known as phorate, which is made in the United States for use overseas, is a common example. It is frequently shipped in 55-gallon drums in sea containers because it is a commercial equivalent to VX nerve agent. It is almost as toxic and has the same immediate effect on humans as does VX.

EXPLOSIVES INCIDENTS

Bomb Squad and HAZMAT Interface

The tasks of bomb squads and HAZMAT teams are combining and in many cases these two specialty groups are working hand in hand. Many bomb technicians are becoming HAZMAT technicians, while many HAZMAT technicians are learning more about the bomb squad duties. One cannot replace the other, but the best approach is a team or coordinated approach to an explosive device that may be chemical in nature. The division of duties is pretty clear: the bomb squads handle anything explosive and the HAZMAT teams handle the chemicals. The problem comes into play when the device combines both of these hazards. Preplanning and training are key to success at one of these events. A disturbing trend, although slight, is the use of chemicals and explosives. The building of an explosive chemical dispersion device is complicated and unless done just right it will not disperse the chemical agent. Just the search of a suspect's home is complicated as both the bomb tech and the HAZMAT tech are looking for two entirely different hazards. When chemicals or explosives are suspected the best approach is to use people from both of these disciplines to make the first search. The level of protective clothing is determined by the hazards presented, but may involve a bomb suit with a chemical suit underneath and a SCBA for respiratory protection.

There are situations in the HAZMAT world in which the tools used by a bomb technician have some applicability. The most common tool would be the remote control robot, which has live video capability and can be specially outfitted with air monitors that use radio telemetry to communicate with a computer to provide air monitor readings and images of a HAZMAT situation. The pan disrupter, or water cannon, is another tool that may be used to open dangerous containers from a remote location. In some special situations a vent and burn may be necessary, which is a job suited for a bomb technician and other tank specialists.

General Rules

Law enforcement's actions are limited at known explosives events, usually to evacuation, isolation, and crime scene tasks. The bomb squad will be called to handle the actual device. Even if a bomb dog did not hit on a package, only a bomb technician should handle the package. Local protocol will dictate how responders respond to a bomb threat, and as each jurisdiction has differing thoughts on the subject it is not discussed here. But if a suspicious package or device is found then responders should evacuate everyone from around the package and isolate the area. The distances used for these types of events are provided in Figure 8-5, however, always consult with the local bomb squad for their preferences. The entry of a bomb technician to examine a suspicious package is equivalent to that of a HAZMAT technician going in to look at a leaking container. Both are risky, but both are very comparable in setup and mitigation. If one were to follow standard HAZMAT practices for a suspected package the end result would be a safe and efficient operation. There is the same need for isolation, identification, protective clothing, monitoring, scene setup, mitigation, and termination. Other responders can assist in the handling of explosives events by following the same protocols and making sure the scene is safe.

SAFETY As tempting as it may be for a responder to touch, open, or move a suspicious package, this job should be left to a bomb technician who will most likely open the package remotely.

Special Situations

One of the major reasons that HAZMAT and bomb technicians are coordinating their efforts is the increase of chemical bombs. Another possibility is that a criminal may use their knowledge of bomb squad operations to their advantage. Much of what a bomb technician does to a suspicious package may be explosive, or at least provide an ignition source, which could potentially level a building if the criminal left a flammable material behind. The remote opening of a package may be just the "booster" charge a large explosive or flammable device needs to be detonated. Another possibility is that the bomb technician through the remote opening of the package becomes the dispersal method for a chemical agent that can then hurt or harm all of the investigators. Keep in mind that the criminal may not want to

CASE STUDY

The bomb squad was requested to a federal facility where a package was thought to contain an explosive device. The federal facility where the package was x-rays all of the mail coming into the facility. The foot locker, which had a legitimate return address, had been overnighted to a specific person within the agency. When x-rayed the security people saw what they thought was an explosive device. The bomb technician looked at the X rays and determined that in addition to bomb-like items there were containers of chemicals present. The X rays showed batteries, wires, jars with chemicals, pipes, and compressed gas cylinders. He requested the assistance of the HAZMAT team because this was a combination device. The decision was made to move the package to a remote location where it could be opened. When it was opened it was determined that junk parts from several motorcycles were in the trunk, which included batteries, wires, mufflers, spray cans, and oil. Following discussions with the security people it was noted that this was not the first time this person had mailed weird packages to the building. His previous packages included cat and dog body parts all mailed to his former supervisor. Since no threats were communicated, there was no crime involved in this junk mail case. The concern is though that when the next package arrives the security personnel might downplay the need for proper safety precautions, and the next package may be the real deal.

hurt or harm anyone just then, instead they may be looking for the harm to show up years later.

There have been several recent events involving chemicals that were either the initiating device or were the target of an explosive charge. A cylinder can be ruptured and land upwards of a mile, and a 55-gallon drum can be ignited and fly several thousand feet. One recent threat involved an explosive device that was to be placed on a liquid propane tank. A household-size propane tank, such as that used for a barbeque grill, if ruptured provides a kill zone from the flame 35 feet in diameter. The target of the propane attack just north of Sacramento was an 8-million-gallon tank of liquid propane. As with the placement of an improvised explosive device (IED) on a cylinder, tank car, or facility tank there is an optimal location to place the device. There are specific methods to ensure detonation, which require specific devices, placements, and weather. If the tank in Sacramento had been detonated the crater would have been several miles in diameter. There was a threat a few years ago by several people who were going to place an IED on a hydrogen sulfide tank to create a diversion so they could rob an armored car. The plot was discovered and arrests were made prior to the device being placed.

A potentially growing threat will be the use of a radiological dispersion device (RDD), which has seen some use overseas. In this country there have been radiological materials used in crimes against individuals, but none involving terrorism. One thought is that the use or threatened use of an RDD will be on the increase, and that it will be used in the place of anthrax hoaxes. The bomb tech and the HAZMAT tech will need to coordinate their efforts to handle this type of threat.

SWAT OPERATIONS

Intel

One important segment to any SWAT or tactical operation is the gathering of intelligence prior to the assault. When chemicals are added to the mixture of hazards, this can further complicate the arrest. It is possible that the chemicals could be used against the arresting officers to thwart the assault. Just being in the proximity of some chemicals may place the SWAT team in harms way. Hopefully through intelligence gathering the presence of chemicals can be discovered and research on the chemical hazards can be done before the entry. If the chemicals are only present in the building then normal SWAT operations can occur to make the arrest. If the chemical could be booby-trapped or used as a protection mechanism then additional planning is necessary. If the assault is going to take place at an industrial location or one that stores or uses chemicals then information may be available for planning purposes. Chapter 1 provides some background on the Emergency Planning and Community Right to Know Act (EPCRA), which requires information about chemical storage at facilities. If the facility is required to report under this act, then there is a good bit of information that is available on the facility, the chemicals, and where and how they are stored. If a suspect has stolen chemicals from a facility then the list of potential materials is also available through this information. Every commercial establishment is required to keep a list of chemicals stored on site, and is supposed to have Material Safety Data Sheets (MSDSs). The type and the quantity of the chemicals determine if this information is distributed elsewhere. If the facility is covered by the OSHA process safety regulation or the EPA Risk Management Program then chemical information on the facility is available for employee, emergency responder, and, in some cases, the public's use. Although critical information on the facility is not supposed to be in public view the facility had to develop a worst-case scenario if a major accident occurred. Part of the development of the worst-case scenario in the facility is to identify the specific steps that could be taken to make the worst-case scenario come true. Although not for public viewing it is available for employees and emergency responders, and it is the "how-to" information to destroy that plant and in some cases cause havoc in the community for many miles. Either the SWAT intelligence component or the terrorism division of the police department should be aware of all the facilities in their area that could have that kind of impact.

> **NOTE** A 90-ton railcar of chlorine if attacked in the proper fashion could create serious effects, including fatalities in a path of about 2000 feet by 9 miles, and could impact up to 24 miles downwind.

Special Considerations

The planning for a SWAT OPS that involves chemicals will have some additional steps to plan for. The basic assault plan will remain the same with contingencies in place to deal with the chemicals. Many of the issues with the chemicals will be implemented only if the chemicals leak or are caused to be released. The biggest threat in these operations will always be the two-legged threat, with a bullet being the primary killing agent. When dealing with chemicals, even those set to be released on the SWAT operators, the mission is such that the entry can be made and the subject apprehended, but that a quick exit will be necessary to make the operation safe. There are some SWAT teams across the country practicing in Level A chemical protective clothing to perform their assaults. There is nothing stealthy about wearing a Level A suit, it is big, bulky, noisy, and usually a bright neon color, and the heat stress is tremendous. It is nearly impossible to shoot wearing this type of garment. The practical side of SWAT entries has the SWAT operators wearing military-style protective clothing such as the Saratoga gear. The key to a SWAT operator's protection is respiratory, and the military version has the SWAT operator wearing a

Powered Air Purifying Respirator (PAPR), which is an APR with an air pump attachment. Many SWAT teams use a gas mask (also known as an APR) when making entries. These masks require the wearer to inhale deeply with some effort to get a breath of air, while a PAPR provides a steady stream of cool air across the face, which can be easily inhaled. When making an entry into a building that may have chemicals, the SWAT operator should have some type of respiratory protection available. When it is likely that the chemicals are going to be used as a protection mechanism then the SWAT operator should be wearing military chemical gear and a PAPR. When higher levels of protective clothing are required it may be best to wait out the suspect or use other means. The best clue as to the safety of a building is to watch the suspect. If the suspect is walking around in a Level A chemical suit, then you may want to wait prior to entry. If the suspect isn't showing any signs or symptoms, then the airborne concentration of the chemical must be low. SWAT operators should have protective clothing in the event that the area is booby-trapped or during the arrest something gets knocked over or spilled. The use of Nomex should be at a minimum, since the fire hazard will probably be the most likely hazard for an operator to encounter. In many cases entries are made into drug labs or other chemical areas using chemical protective clothing, and the biggest risk to the entry person is the fire risk. Chemical protective clothing and heat do not mix, you could instantly become shrink-wrapped. The determination of protective clothing is based on the type of hazard present. Chapter 4 discusses standard protective clothing, and Chapter 5 discusses rescue operations. The SWAT operator is performing the same type of operation as a firefighter making a rescue, and the risks are going to be comparable. The biggest change is that the person the SWAT operator is going after has made a conscious decision to be in that environment, and when endangered the SWAT operator should wait out the suspect. The presence of hostages obviously changes the picture and makes it much more complicated.

> **SAFETY** SWAT operators must not let the chemicals get in the way of their operation, they must remain focused on their primary mission.

> **SAFETY** Training with the gear, especially shooting, is essential for proficiency.

PROTECTIVE CLOTHING AND THE SWAT ENVIRONMENT

A SWAT operator should always be able to wear their body armor, and the chemical protection be secondary to gunshot protection. A SWAT operator must be able to see and be able to pull the trigger with a minimum of distraction. The best protection that does offer chemical protection is military gear, known as **Mission Oriented Protective Posture (MOPP) gear.** Designed for the military environment, which has some crossover into the civilian SWAT world the military gear offers protection against chemical exposures. The nature of the SWAT entry means that the actual contact time with any chemical will be short. Exceptions to this thought would be a booby trap that covers a SWAT operator or an operator who falls into a chemical release. The military gear is designed for military warfare chemicals and is to be used for long periods of time in a contaminated environment. The SWAT environment is a quick-in/quick-out and hopefully minimizing contact with any chemicals. The standard military issue clothing is the Saratoga Joint Service Lightweight integrated systems technology (JSLIST) suit designed for chemical and biological warfare. It is accompanied by the MCU2P gas mask, which is an air purifying respirator (APR) as described in Chapter 4. The Saratoga comes with a SCALP suit, which is an overgarment used for liquid splashes, and it also comes with chem/bio underwear. The mask, which is called the M40 mask in the military world, can only be used when the oxygen level is above 19.5 percent for non-IDLH atmospheres and nonflammable atmospheres. The cartridge filter is for military chemicals. There are other canisters available for industrial chemicals that may be better suited for a wider range of chemicals. The C2 filter that comes with the mask will protect against military chemical and biological materials, acid gases (hydrogen chloride, chlorine, hydrogen fluoride, sulfur dioxide, sulfuric acid, and nitric acid), fuels, oils, and alpha and beta radiation. It does not protect against ammonia, carbon monoxide or carbon dioxide, nitric oxide, nitrogen dioxide, or metal carbonyls. The basis for the protection for other materials is based on the vapor pressure of the chemical. The gear offers protection for chemicals with a vapor pressure of less than 7.5 mm Hg, and offers limited protection for vapor pressures between 7.5 and 75 mm Hg. The higher the concentration the less time that can be spent in that environment, the lower the concentration the more time that can be spent. Chapter 2 provides more information on vapor pressures. The biggest challenge to this gear would be an ammonia environment, however, a commercial cartridge is readily available to protect against the ammonia. The glove issue is one that is controversial, but the SWAT operator should not be touching any open chemicals, thus the hand protection should allow the operator to shoot well. A thin nitrile or surgical glove with a Nomex flight glove overcover is adequate protection for a SWAT operator. If chemical contact or immersion is a real possibility then the hand protection can be increased. As long as there are live people without protective clothing in the environment, no matter what happens, with the exception of an explosion or rapid fire spread, the Saratoga gear along with Nomex jumpsuits would be the preferred gear for SWAT operations. While using this gear you lose vision, communication, stealth ability, quickness, and you gain heat stress.

Setup

When hazardous materials are involved in a SWAT operation there are a few extra steps that need to be added to the planning process. Several contingencies need to be added to the process so that in the event of a release operators are prepared. When chemicals are known to be involved, someone must be assigned to research the materials and learn about their hazards. Storage information, typical containers, basic hazards, and the type of PPE required are all essential information. The operators must be briefed on what to look for, what type of odors may be present, and what conditions would trigger a retreat situation.

The entry into the building may change slightly, especially if explosive breaching is going to be done to gain entry. If flash bangs are going to be authorized then some additional checks will have to be done. Before detonating a **breach charge** the breacher or someone next in line should check the door area for at least 5 seconds with a multigas monitor as well as a photoionization detector, which is shown in Figure 8-6. If the bad guy is sure that breaching or flash bangs are going to be used, then a flammable atmosphere could be created and the SWAT operators provide the ignition source to launch themselves and the building a considerable

Figure 8-6 Prior to entering the building a SWAT operator is checking the atmosphere near the door. If conditions are ok, when the green light is given to initiate the entry, and in some cases the team may need to withdraw.

distance. Most air monitors can be modified to vibrate when in alarm, and the operator should not concentrate on the instrument. They should make the initial check, then fall into place, make the assault, and capture the suspect. If the meter alarms, the operator should evaluate the alarm, determine the risk, and in most cases complete the mission, and then with the suspect evacuate the building. When operating in environments that are likely to have chemical contamination or very toxic materials, the operators should don respiratory protection prior to entry. If they choose not to use respiratory protection, then an alarm on the meter will mean that respiratory protection is needed.

Another consideration is decontamination of the operators, hostages, and the suspect. The best group to perform this decontamination is a group of HAZMAT-trained law enforcement officers, who could stand by prior to the assault. The second-best group to perform this task is a HAZMAT team, who could be standing by prior to the assault. Obviously **operational security** is key prior to an assault, so having a good relationship with the local HAZMAT team is crucial. Unfortunately, a large-scale decon set up takes time to set up, which means preparation time is crucial. The local HAZMAT team can provide some guidance based on the chemicals on some decon methods that could be employed while a decon setup is being deployed. Any SWAT operation whether chemicals are thought to be present or not should have a contingency for decon. If a suspect throws a jar of sulfuric acid at an officer entering the suspect's bedroom, there should be no

hesitation in moving that officer to a location where the officer should be immediately decontaminated. A shower, a hose, or even a simple garden sprayer are all adequate forms of emergency decontamination. The commander must always have a plan to deal with such emergencies, and have personnel assigned to complete those tasks.

Emergency Conditions

The SWAT team enters the door of a house and as soon as they enter the house they find a liquid on the floor. If they are wearing the Saratoga gear with respiratory protection they can continue on to complete the mission. If they are not wearing Saratoga gear or not wearing respiratory protection then a decision must be made as to continue or to evacuate. That decision has to be made with an air monitor to determine if the risk is fire, corrosive, or toxic hazards. No matter the situation, decontamination is crucial because some of the operators will have direct contact with the product. Those who did not come in contact do not need to be decontaminated, but do require a medical evaluation depending on the type of chemical found. There are a number of ways for a SWAT team to be set up, and for a criminal to use known entry procedures against the SWAT team. Some criminals may inadvertently have chemicals in their proximity and the SWAT team through a normal operation ignites them or causes a release, as shown in Figure 8-7. Imagine a bank robber who is using acetone to wash the dye pack ink from the money and has the money and acetone in a bucket. The SWAT team enters the building and throws a flash bang near the bucket. Acetone evaporates quickly, is heavier than air, is highly flammable (explosive), and is readily ignited. The resulting explosion could blow the SWAT team across the street and severely damage the building, injuring or killing those in or near the room. The same change of mind-set has to occur with SWAT operators as happened with the bomb technicians.

SAFETY Hazardous materials present some risks in normal handling, however, in the hands of a criminal they may have disastrous effects.

When planning an assault there should be operators assigned to handle the air monitor(s), operators assigned to check for hazardous materials, as shown in Figure 8-6 and Figure 8-7, along with other normal checks, operators assigned to perform rescues, and some assigned to do decontamination. These operators do not have to stand by

Figure 8-7 SWAT teams should not alter their basic assault plan, just because chemicals may be present. The biggest threat comes from the criminal inside, and the chemicals are a secondary threat. If leaking materials are encountered then they should e checked for fire, corrosive or toxic hazards, after the building has been cleared of the human threat.

Figure 8-8 An example of a drum dump, in which the investigators are continuing on with the sampling plan that was developed on the previous entry. Drums must be sampled and characterized, and evidence collected.

while the others make the entry, they can be part of the team. In the event of a contamination, they then switch to these collateral duties as long as the human threat is removed. If intelligence reveals that chemical contamination is possible, then other personnel may need to be added to the assault group, possibly staged away from the event. The local HAZMAT team could be requested to stand by at a nearby location. This can be accomplished without operational security being compromised. Many large-city HAZMAT teams are used as standby details for other search warrants and are provided a time and a location for the standby, with no other details being provided. Knowing the type of chemicals involved and the expected tasks to be performed will speed up the HAZMAT team's response once notified. Liaisons should be established prior to an emergency so that relationships and work tasks are predefined, making the operation safer and more efficient.

EVIDENCE COLLECTION

Considerations

The collection of evidence in a hazardous environment brings a whole host of challenges. Luckily in most cases the collection of evidence is not time dependent and fingerprinting can often wait until the HAZMAT hazard is reduced or removed. There are cases though when time is of the essence and when it needs to be determined if the environment is safe to collect the evidence. In many cases protective clothing must be worn to determine the nature of the hazard and to determine what type of equipment must be used to collect the evidence, much like the **environmental crime** shown in Figure 8-8.

The considerations for evidence collection are multiple. First and foremost the persons collecting the evidence must be trained to a level of HAZMAT training. This level is determined by the tasks that are going to be performed by the investigator. If the investigator is collecting normal evidence with chemicals present or will be taking chemical samples from open sources then the operations level of training is required. If the investigator is going to be specifically collecting chemical evidence or is going to be opening containers then the technician level is required. Both the operations courses and the technician courses should be geared toward law enforcement tasks, and the operations course specifically should cover standard HAZMAT operations issues and needs to add some specific chemical sampling training to meet OSHA requirements. The tiered approach works well in that operations-level personnel can evaluate the hazards, and if necessary request additional resources as required. There are not many law enforcement departments that have officers trained to the HAZMAT technician level, thus they may need to rely on the local HAZMAT team for this level. The local HAZMAT team can assist and offer guidance, but with some exceptions most HAZMAT teams are not trained in evidence collection. The best approach is to combine the two groups and make a combined entry, with both focusing on their specific tasks. Other possible resources are the bomb technicians, since most are becoming HAZMAT technicians, and the fire investigators, since to investigate a chemical fire they are required to be HAZMAT technicians. The EPA offers evidence collection courses geared toward hazardous materials and the local EPA contact is a good resource. The local EPA criminal investigator is also a good contact, since most of their investigative work involves the collection of evidence in hazardous atmospheres. The Coast Guard may also have investigators who can provide assistance in this area. Another potential source of assistance is the local or state environmental crimes investigators who are also trained in this specialty.

Other considerations for evidence collection are the sampling media, tools, and equipment. Each of these materials must be compatible with the chemical so as not to cause a reaction or change in the chemical makeup of the sample. The protective clothing that is used must protect the wearer from the type of hazards that are present. Some form of respiratory protection may be required, which has time limitations and may have other conditions of use especially when using APRs and PAPRs. The time that personnel can collect evidence is another issue, as crews must be sent in teams of two with a backup team of at least two. If using SCBA then the time is limited to a maximum of 30 to 40 minutes, if not shorter. Crews must be rotated in because the PPE will wear them down and rehabilitation and hydration are vital.

Evidence must be collected in compatible containers, and how soon the material needs to get to a lab must be considered. Will the local lab accept this type of material and are they equipped to store and analyze the material? Labs will have certain requirements for storage containers and may have requirements for certain types of collection media. An important question to ask is how much of a sample is required. Discussions need to take place with the lead investigator and the prosecutor as to the amount of chemicals that are needed for evidence and for prosecution. If the suspect had a gallon jar of a flammable substance, does the whole jar need to be kept or can a photograph and a 40-ml sample suffice for evidentiary purposes?

Operations and Entry

Operationally, the setup will be very similar to a HAZMAT accident and will require the same sectoring and operational concerns. The complicating factor in collecting evidence in a HAZMAT situation is that **chain of custody** must be maintained. The actual sectoring of the event will make tracking personnel easier, and access points are usually well controlled at a crime scene and may need to be strengthened when chemicals are involved. Safe work practices need to be employed so that cross-contamination does not occur and that evidence is collected in a manner that cannot be disputed in court.

SAFETY When chemicals are discovered at a crime scene or are known to be present, the first priority is life safety.

Persons in and around the area should be evacuated until the nature of the hazard is determined. This may mean a single apartment or the entire apartment building. The chemical determines the actual size of the isolation zone. Reference books and air monitors can assist in this decision. The ultimate decision needs to be made with an air monitor as that provides real time on scene conditions. Prior to entry the criminal threat should have been removed and should not be an issue. Initial entry crews should enter the building in appropriate protective clothing and accompanied by air monitors that protect them against fire, corrosive, and toxic hazards. A minimum backup crew of two is required, and someone outside must be designated as the Incident Commander. If the event is large enough a separate safety officer must be assigned, if not the Incident Commander assumes that responsibility. If available, assisting in this entry should be a HAZMAT technician, and if there is any chance that explosives or explosive-making materials may be present, a bomb technician. The hazards should be characterized and leaking or open containers noted and possibly mitigated. A basic drawing that will be used to develop a **sampling plan** should be completed and

Polaroid photographs should be taken so that other crews can be briefed prior to entry, as is shown in Figure 8-2. Although all of this information is discoverable, crews are not yet collecting evidence because this recon is to assist in the sampling plan, which will provide subsequent crews with the knowledge of where to begin the evidence collection. Decontamination must be planned for, although unless the initial entry crew steps in something or comes in contact with a solid or a liquid they should not require much, if any, decon. Other crews who will be sampling chemicals will require some decon, but this may be accomplished with dry decon or a bucket to clean their hands. Contingency plans should be developed in the event of a fire, massive release, failure of a container, or gross contamination of an investigator.

An investigator should be assigned to track the entry team, to monitor the time on air, or the work time, and to possibly coordinate the evidence coming from the hot sector. There are a lot of logistical concerns and an investigator should be assigned to logistics to make sure all of the equipment is readily available. If required the decon crews usually take a beating, since they are dressed in PPE for long periods of time and are usually not switched out as the entry crews are switched. They should be monitored by the safety officer, and someone should be assigned to monitor these crews.

The initial entry crew in addition to developing the concerns for the development of a sampling plan should also be determining the need for PPE. If they mitigate some hazards or ventilate, the level of PPE may be able to be downgraded for the next entry. The level of PPE is determined by the type of chemical and the risk that is present. If the chemical released presents a fire risk then fire-resistive clothing is required such as a Nomex or PBI jumpsuit or firefighters' turnout clothing. Crews should not be entering a flammable environment with chemical protective clothing. The chemical has to be removed or the building ventilated so as to remove the fire hazard. When dealing with chemicals respiratory protection is always recommended, but the least amount of protective clothing is better for the entry crews. The other factors such as heat stress, mobility, vision, and hearing also need to be considered in the hazard assessment. Conditions should be reevaluated on a regular basis to make sure conditions have not changed.

SUMMARY

The addition of chemical hazards in the law enforcement field complicates an already complex task. A criminal with bad intentions could wreak havoc with law enforcement officers using chemicals, and officers must be prepared for this possibility. Terrorism has forced HAZMAT teams to become better at their present jobs, and hopefully this trend also moves the law enforcement community to make the same transition. On the federal level several law enforcement groups have been providing HAZMAT training for their investigators, much like the HAZMAT teams across the country. This trend needs to move to the local level for a number of reasons: first, the law requires it and second, most importantly, it is for your protection.

The unfortunate thing is that some chemicals do not cause noticeable or immediate effects, but instead they may affect your family later or may even kill you five or ten years later.

SAFETY Going into a crime scene ill prepared for chemicals is asking to get shot by a different means.

KEY TERMS

Biological weapons lab   A lab designed to manufacture biological warfare agents.

Breach charge   An explosive charge such as a strip or shape charge that is used to explosively remove a door or window.

Chain of custody   All evidence must be tracked and must have a written description of how it moved from the crime scene to the lab and then to storage. Each time it passes from one person to another it must be documented.

Chemical weapons lab   A lab designed to manufacture chemical warfare agents. This type of lab is either manufacturing the agent from precursor chemicals or is synthesizing commercially available pesticides.

Environmental crime   The spilling, release, or potential release of chemicals into the environment. Littering, discharging chemicals into a sewer, creek, or other waterway is an environmental crime.

Explosives lab   A lab designed to manufacture explosives.

Huffing   Term used to describe the intentional inhalation of paints, solvents, or other chemicals with the intention of getting high.

Mission Oriented Protection Posture (MOPP) gear   Chemical-resistant clothing designed for military operations.

Operational security   Keeping the details of an operation secret and only divulging critical information to those who have a need to know.

Powered Air Purifying Respirator (PAPR)   An air purifying respirator (APR) that has a blower attachment that blows filtered air to a mask.

Sampling plan   A drawing or a map of an incident that indicates all of the containers that need to be sampled. A priority of containers to be sampled and the resources necessary to do that sampling are part of the sampling plan.

Universal precautions   The wearing of PPE designed for blood-borne pathogen protection, such as gloves, gown, and a face mask.

REVIEW QUESTIONS

1. What types of HAZMAT situations are common to law enforcement officers?
2. When should universal precautions be used?
3. What are the three types of labs that may be encountered?
4. Intelligence reveals that chemicals may be used to booby-trap a building, and the SWAT team is developing a plan of action. What special considerations need to be added for this plan?
5. What are the minimum evacuation distances for pipe, car, van, and truck bombs?
6. What are some additional evidentiary considerations for evidence collection in hazardous environments?

Suggested Readings

Hawley, Chris, *Air Monitoring and Detection Devices,* Delmar, a division of Thomson Learning, Albany, NY, 2001.

Lesak, David, *Hazardous Materials Strategies and Tactics,* Prentice Hall, Upper Saddle River, NJ, 1998.

Noll, Gregory, Michael Hildebrand, and James Yvorra, *Hazardous Materials: Managing the Incident,* Fire Protection Publications, Oklahoma University, 1995.

Schnepp, Rob, and Paul Gantt, *Hazardous Materials: Regulations, Response, and Site Operations,* Delmar, a division of Thomson Learning, Albany, NY, 1999.

GLOSSARY

Absorption   Defensive method of controlling a spill by applying a material that absorbs the spilled chemical.

Acute   Quick, one-time exposure to a chemical.

Air Bill   Term used to describe the shipping papers used in air transportation.

Anthrax   Naturally occurring biological material that is severely toxic to humans; common agent cited in hoax incidents.

Aboveground storage tanks (AST)   Tanks that are stored above the ground in a horizontal or vertical position. Smaller quantities of fuels are often stored in this fashion.

Air purifying respirators (APR)   Respiratory protection that filters out contaminants out of the air, using filter cartridges; requires the atmosphere have sufficient oxygen along with other regulatory requirements.

Awareness   The basic level of training for emergency response to a chemical accident, the ability to recognize a hazardous situation and call for assistance.

Biological agents   Toxic substances that are living materials or are obtained from living organisms.

Biomimetic   Form of a gas sensor that is used to determine levels of carbon monoxide. Typically used in residential CO detectors, it closely recreates the body's reaction to CO and activates an alarm.

BLEVE   Boiling liquid expanding vapor explosion, which occurs when a pressurized tank is heated and the pressure inside the tank exceeds the capacity of the relief valves to handle the pressure. When pressurized tanks rupture, they can travel significant distances, and if holding flammable materials, they will usually ignite, resulting in a large fireball.

Blister agents   See *vesicants*.

Blood agents   Chemicals that affect the body's ability to use oxygen and in many cases prevent the body from using oxygen, producing fatal results.

Boiling point   Temperature to which liquids must be heated to turn to a gas.

Building Officials Conference Association (BOCA)   A group that establishes minimum building and fire safety standards.

Bulk tank   Large transportable tank comparable to but larger than a tote.

Butyric acid   Fairly common lab acid used in many attacks on abortion clinics; although not extremely hazardous, it has a characteristic stench that permeates the entire area where it is spilled.

Calibration   Setting the air monitor to read correctly. To calibrate a monitor, the user exposes it to a known quantity of gas and makes sure it reads the values correctly.

Carcinogen   Material capable of causing cancer in humans.

Catalytic bead   The most common type of combustible gas sensor, which uses two heated beads of metal to detect the presence of flammable gases.

Ceiling level   Highest exposure a person can receive without suffering any ill effects; combined with the PEL, TLV, or REL, it establishes a maximum exposure.

Chemtrec   The Chemical Transportation Emergency Center, which provides technical assistance and guidance in the event of a chemical emergency. This network of chemical manufacturers provides emergency information and response teams if necessary.

Chip measuring system (CMS)   Form of colorimetric air sampling in which the gas sample passes through a tube. If the correct color change occurs, the monitor interprets the amount of change and indicates the gas level on a LCD screen.

Choking agents   Chemicals that produce an immediate reaction (coughing, difficulty breathing); terrorism agents that are considered choking agents are chlorine and phosgene, both very toxic gases.

Chronic   Continual or repeated exposure to a hazardous material.

Computer Aided Management for Emergency Operations (CAMEO)   Computer program that combines a chemical information database with emergency planning software; commonly used by HAZMAT teams to determine chemical information.

Consist   Shipping papers that list the cargo of a train; the listing is by railcar of all cars.

Cryogenic gas   Any gas that exists as a liquid at a very cold temperature, always below $-150°F$.

Damming   Stopping a body of water, which at the same time stops the spread of the spilled material.

Dangerous Cargo Manifest (DCM)   Shipping papers for a ship, which list the hazardous materials on board.

DECIDE   Management system used to organize the response to a chemical incident. The factors of DECIDE are: detect, estimate, choose, identify, do the best, and evaluate.

Decontamination   Physical removal of contaminants (chemicals) from people, equipment, or the environment; most often used to describe the process of cleaning to remove chemicals from a person.

Deflagrates Rapid burning, which can be considered a slow explosion, but travels at a lesser speed than a detonation.

Department of Transportation (DOT) Federal agency that issues regulations regarding the transportation of hazardous materials.

Diking Defensive method of stopping a spill; a common dike is constructed of dirt or sand and is used to hold a spilled product. In some cases, in a facility a dike may be preconstructed, such as around a tank farm.

Dilution Adding a material to a spill to make it less hazardous; in most cases water is used to dilute a spilled material, although other chemicals could be used.

Diverting Using materials that can move products around or away from an area; for example, several scoops of dirt can be used to divert a running spill around a storm drain.

8-step process© Management system used to organize the response to a chemical incident. The elements are: detect, estimate the harm, choose an objective, identify the action, do the best you can, and evaluate your progress.

Emergency decon Rapid removal of material from a person, who requires immediate cleaning. Most emergency decon setups use a single hoseline to perform a quick, gross decon.

Emergency Planning and Community Right to Know (EPCRA) The portion of SARA that specifically outlines how industries report chemical inventory to the community.

Emergency response planning (ERP) Levels that are used for planning purposes and are usually associated with preplanning for evacuation zones.

Encapsulated suit Chemical suit that covers the responder, including the breathing apparatus; usually associated with gastight and liquid-tight Level A suits, but some Level B styles are fully encapsulated but not gas or liquid tight.

Endothermic reaction Chemical reaction in which heat is absorbed and the resulting mixture is cold.

Environmental Protection Agency (EPA) Federal agency ensuring protection of the environment and the nation's citizens.

Etiological Hazard that includes biological agents, viruses, and other disease-causing materials.

Evacuation Movement of people from an area, usually their homes, to another area that is considered to be safe. Persons are evacuated when they are no longer safe in their current area.

Exothermic reaction Chemical reaction that releases heat, as in when two chemicals are mixed and the resulting mixture is hot.

External floating roof tank Tank in which the roof is exposed to the outside and covers the liquid within the tank. The roof floats on the top of the liquid, which does not allow for vapor build-up.

Extremely Hazardous Substances (EHS) List of 366 substances that the EPA has determined present an extreme risk to the community if released.

Fine decon More detailed form of decontamination, usually performed at a hospital by staff who are trained and equipped to perform decon procedures.

First responders A group designated by the community as those who may be the first to arrive at a chemical incident. This group is usually comprised of police officers, EMS providers, and firefighters.

Fit testing Test that ensures that the respiratory protection fits the face and offers maximum protection.

Flammable range The correct mixture of flammable gas and air in which there can be a fire; for most gases there is a certain span (range) that will allow for a fire to ignite.

Flash point Temperature of a liquid, which if an ignition source is present, will ignite only the vapors being produced by the liquid, creating a flash fire.

Formal decon Washing and scrubbing portion of the decontamination process. The process is usually repeated and is performed by a decon crew.

Frangible disk A pressure-relieving device that ruptures to vent the excess pressure. Once opened, the disk remains open; it does not close after the pressure is released.

Freezing point Temperature at which liquids become solids.

Gas State of matter that describes the material in a form that moves freely about and is difficult to control. Steam is an example.

GEDAPER© Management system used to organize the response to a chemical incident. The factors are: gather information, estimate potential, determine goals, assess tactical options, plan, evaluate, and review.

Gross decon Step in the decontamination process that removes the majority of the chemicals through a flushing of the person. The gross washing employs large amounts of water and is usually done by the individual or a partner.

Gross negligence Disregarding training and acting without regard for others.

Hazardous materials (HAZMAT) Those substances that have the ability to harm humans, property, or the environment.

Hazardous Waste Operations and Emergency Response (HAZWOPER) The OSHA regulation that covers safety and health issues at hazardous waste sites and response to chemical incidents.

Hyperbaric chamber Chamber used to treat scuba divers who ascend too quickly and need extra oxygen to survive. The chamber, which recreates the high pressure atmosphere of diving and forces oxygen into the body, is also used to treat carbon monoxide poisonings and smoke inhalation.

ICt_{50} Military term for incapacitating level over set time to 50 percent of the exposed population.

Ignition temperature   Temperature of a liquid that will ignite on its own without an ignition source; also called *self-accelerating decomposition temperature.*

Incendiary agents   Chemicals used to start fires, the most common being a Molotov cocktail.

Incident Commander   HAZMAT training level that encompasses the operations level with the addition of incident command training. Intended to be the person who may command a chemical incident.

Infrared sensor   Monitor that uses infrared light to determine the presence of flammable gases. Light is emitted in the sensor housing and the gas passes through the light; if it is flammable, the sensor indicates the presence of the gas.

Intermodal containers   Constructed to be transported by highway, rail, or ship. Intermodal containers exist for solids, liquids, and gases.

Internal floating roof tank   Tank with a roof that floats on the surface of the stored liquid; this tank also has a cover on its top to protect the top of the floating roof top.

Irritant   Material that is irritating to humans, but usually does not cause any long-term effects.

Isolation area   Area in which citizens and responders are not allowed; may later become the hot zone/sector as the incident evolves. This is the minimum area that should be established at any chemical spill.

Laws   Legislation approved by Congress and signed by the president.

LCt_{50} Military term for lethal concentration over set time to 50 percent of the population.

Leaking Underground Storage Tank (LUST)   Term to describe this potential health and environmental hazard.

Level A protective clothing   Fully encapsulated chemical protective clothing; gastight and liquid-tight gear offers protection against chemical attack.

Level B   Level of protective clothing usually associated with splash protection; Level B requires the use of SCBA.

Liability   Being held legally responsible for your actions.

Liquid   State of matter that implies fluidity, which means a chemical can move as water would. There are varying states of being a liquid from moving very quickly to very slowly. Water is an example.

Local Emergency Planning Committee (LEPC)   Group comprised of members of the community, industry, and emergency responders to plan for a chemical incident and to ensure that local resources are adequate to handle an incident.

Lower explosive limit   The lower part of the flammable range, or the minimum required to have a fire or explosion.

Mass casualty   Incident in which the number of patients exceeds the capability of the EMS system to manage the incident effectively. In some jurisdictions this can be two patients, while in others ten may qualify as a mass casualty.

Material Safety Data Sheets (MSDS)   Information sheets for employees that provide specific information about a chemical, with attention to health effects, handling, and emergency procedures.

Melting point   Temperature at which solids become liquids.

Metal oxide sensor (MOS)   Coiled piece of wire that is heated to detect the presence of flammable gases.

Multi-gas detector   Air monitor that measures oxygen levels, explosive (flammable) levels, and one or two toxic gases, such as carbon monoxide or hydrogen sulfide.

National Fire Protection Association (NFPA)   A group that issues fire and safety standards for industry and emergency responders.

National Response Center (NRC)   Location a spiller is required to call to report a spill if it is in excess of the reportable quantity.

Negligence   Acting in an irresponsible manner, or different from the way one was taught to act, differing from the standard of care.

North American Emergency Response Guidebook (NAERG)   Reference book provided by the DOT to assist first responders in making decisions at a transportation-related chemical incident.

Occupational Safety and Health Agency (OSHA)   Federal agency responsible for enacting safety regulations protecting the nation's workers.

Operations   Mid level of HAZMAT training above awareness; provides the foundation that allows for the responder to perform defensive activities at a chemical incident.

Ordinary tank   Horizontal or vertical tank that usually contains combustible or other less hazardous chemicals. Flammable materials and other hazardous chemicals may be stored in smaller quantities in these types of tanks.

Overpacked   Response action that involves placing a leaking drum (or container) into another drum. Some drums are made specifically to be used as overpack drums and are oversized to handle a normal-sized drum.

Paragraph q   Section within HAZWOPER that outlines the regulations governing emergency response to chemical incidents.

PDD 39   Presidential Decision Directive 39, which established the FBI as the lead agency in terrorism incidents; the FBI is responsible for crisis management; and FEMA is the lead for consequence management.

Permeation   Movement of chemicals through chemical protective clothing on a molecular level; does not cause visual damage to the clothing.

Permissible exposure limit (PEL)   OSHA value that regulates the amount of a chemical that a person can be exposed to during an eight-hour day.

Personal protective equipment (PPE)   Equipment and clothing designed to protect a person from a variety of hazards when responding to a HAZMAT incident; PPE ranges from earplugs to a fully enclosed chemical suit.

Photoionization detector (PID)   Air monitoring device used by HAZMAT teams to determine the amount of toxic materials in the air.

Polymerization   A runaway chain reaction that can be violent if contained. Once started it cannot be stopped until the chemical reaction has run its course.

Radiological dispersion device (RDD)   Radioactive material that is coupled with an explosive device, which spreads the radioactive material. It is not a nuclear detonation, as the explosion is caused by a standard explosive charge.

Recommended exposure limit (REL)   Exposure value established by NIOSH for a ten-hour day, forty-hour workweek. Is similar to the PEL and TLV.

Regulations   Rules developed and issued by a governmental agency, which have the weight of law.

Relief valve   Device designed to vent pressure in a tank, so that the tank itself does not rupture due to an increase in pressure. In most cases these devices are spring loaded so that when the pressure decreases, the valve closes to keep the chemical inside the tank.

Remote shutoffs   Valves that can be used to shut off the flow of a chemical; the term *remote* is used to denote valves that are away from the spill.

Reportable quantity (RQ)   A term used by the EPA and DOT to describe a quantity of chemicals that may require some type of action, such as reporting an inventory or reporting an accident involving a certain amount of chemicals.

Retention   Digging a hole to collect a spill; can be used to contain a running spill or collect a spill from the water.

Ricin   Biological toxin that can be used by a terrorist or other person attempting to kill or injure someone; the easiest terrorist agent to produce and one of the most common.

Risk-based response   An approach to responding to a chemical incident by categorizing a chemical into a fire, corrosive, or toxic risk. Using a risk-based approach can assist the responder in making tactical, evacuation, and PPE decisions.

Sea containers   Shipping boxes designed to be stacked on a ship and then placed onto a truck or railcar; also called sea boxes.

Sector   See *zone.*

Self-accelerating decomposition temperature (SADT)   See *ignition temperature.*

Self-contained breathing apparatus (SCBA)   Protective gear that safeguards responders' respiratory systems; considered essential for responding to chemical spills.

Sensitizer   Chemical that after repeated exposures may cause an allergic effect for some people.

Shearing effect   Tearing of the molecules, which can be equated to being ripped apart.

Shelter in place   Form of isolation that provides a level of protection while leaving people in place, usually in their homes. People are usually sheltered in place when they may be placed in further danger by evacuation.

Short-term exposure limit (STEL)   Fifteen-minute exposure to a chemical; requires a one-hour break between exposures and is only allowed four times a day.

Sick building chemical   Substances that cause health problems for occupants of a structure.

Sick building   Term associated with indoor air quality; occupants in a sick building become ill as a result of chemicals in and around the structure.

Solid   State of matter that describes chemicals that may exist in chunks, blocks, chips, crystals, powders, dusts, and other types. Ice is an example.

Specialist   HAZMAT training level that provides for a specific type of training, such as railcar specialist; a responder with a higher level of training than a technician.

Specific gravity   Value that determines if the liquid will sink or float in water. Water is given a value of 1, and chemicals with a specific gravity of less than 1 will float on water. Those chemicals with a specific gravity of greater than 1 will sink in water.

Specification (spec) plates   Label on trucks and tanks that lists the type of vehicle, capacity, construction, and testing information.

Standard of Care   Combination of training, experience, standards, and regulations that guide a responder to a specific action.

Standard Transportation Commodity Code (STCC)   Number assigned to chemicals that travel by rail.

Standards   Usually developed by consensus groups establishing a recommended practice or standard to follow.

State Emergency Response Committee (SERC)   Group that ensures the state has adequate training and resources to respond to a chemical incident.

States of matter   Term applied to chemicals to describe the forms in which they may exist: solids, liquids, or gases.

Sublimation   Ability of a solid to go to the gas phase without being liquid.

Superfund Amendments and Reauthorization Act (SARA) Law that regulates a number of environmental issues, predominately for the chemical inventory reporting by industry to the local community.

Supplied air respirators (SAR) Respiratory protection that provides a face mask, air hose connected to a large air supply, and an escape bottle; typically used for waste sites or confined spaces.

Technician High level of HAZMAT training that allows specific offensive activities to take place to stop or handle a chemical incident.

Threshold limit value (TLV) Exposure value that is similar to the PEL but is issued by the ACGIH. It is for an eight-hour day.

Tote Large tank, usually 250 to 500 gallons, constructed to be transported to a facility and dropped for use.

Toxins Substances that are biological in nature and are obtained from naturally occurring materials.

TRACEM Acronym for the types of hazards that may exist at a chemical incident: thermal, radiation, asphyxiation, chemical, etiological, and mechanical.

Underground storage tanks (UST) Tanks, commonly for gasoline and other fuels, that are buried under the ground.

Upper explosive limit (UEL) Upper part of the flammable range; above the UEL, fire or an explosion cannot occur as there is too much fuel and not enough oxygen.

Vapor density Value that determines if the gas will sink or rise in air. Air is given a value of 1, and chemicals with a vapor density less than 1 will rise in air. Those with a density of greater than 1 will sink and stay low to the ground.

Vapor dispersion Intentional movement of vapors to another area, usually by the use of master streams or hoselines.

Vapor pressure Amount of force pushing vapors from a liquid; the higher the force, the more vapors (gas) are being put into the air.

Vapor suppression Defensive strategy to minimize the production of vapors. For example, flammable liquids produce hazardous vapors that could ignite; firefighting foams are used to suppress the vapors and not allow them to move by forming a barrier on top of the spill.

Vesicants Group of chemical agents that cause blistering and irritation of the skin; commonly referred to as *blister agents*.

Waybill Term that may be used in conjunction with consist, but is a description of what is on a specific railcar.

Wheatstone bridge sensor Type of combustible gas sensor that uses a heated coil of wire to detect the presence of flammable gases.

Zone Area established and identified for a specific reason; typically a hazard exists within the zone. Zones are usually referred to as a hot, warm, and cold to provide an indication of the expected hazard in each zone. Some agencies use the term *sector*.

ABBREVIATIONS

ACGIH   American Conference of Governmental Industrial Hygienists

ACS   American Chemical Society

ALK   Alkalis

ALOHA   Aerial Location of Hazardous Atmospheres

ANFO   Ammonium nitrate and fuel oil mixture

APR   Air purifying respirator

AST   Aboveground storage tank

ASTM   American Society for Testing and Materials

ATF   Bureau of Alcohol, Tobacco, and Firearms

BBP   Blood-borne pathogens

BLEVE   Boiling liquid expanding vapor explosion

BNICE   Biological, nuclear, incendiary, chemical, and explosive agents

BOCA   Building Officials Conference Association

CAAA   Clean Air Act Amendments

CAMEO   Computer Aided Management for Emergency Operations

Canutec   Canadian Transportation Emergency Center

CAS   Chemical Abstract Service

CBIRF   Chemical and Biological Incident Response Force (U.S. Marines)

CGI   Combustible gas indicator

Chemtrec   Chemical Transportation Emergency Center

CMS   Chip measurement system

CORR   Corrosives

DCM   Dangerous Cargo Manifest

DNR   Department of Natural Resources

DOT   Department of Transportation

EHS   Extremely hazardous substance

EPA   Environmental Protection Agency

EPCRA   Emergency Planning and Community Right to Know Act

ERP   Emergency response planning

FEMA   Federal Emergency Management Agency

HAZMAT   Hazardous materials

HAZWOPER   Hazardous Waste Operations and Emergency Response

HCS   Hazard communication standard

HMIG   Hazardous Materials Information Guide

HMIS   Hazardous Materials Incident Reporting System

HMRU   Hazardous Materials Response Unit (FBI)

IDLH   Immediately dangerous to life or health

IM   Intermodal

IMS   Incident management system

LC   Lethal concentration

LD   Lethal dose

LEL   Lower explosive limit

LEPC   Local Emergency Planning Committee

LOX   Liquid oxygen

LSA   Low specific activity

LUST   Leaking underground storage tank

MCI   Mass casualty incident

MOS   Metal oxide sensor

MSDS   Material Safety Data Sheet

NA   North American identification number

NAERG   North American Emergency Response Guidebook

NBC   Nuclear, biological, and chemical agents

NDPO   National Domestic Preparedness Office (FBI)

NFPA   National Fire Protection Association

NIOSH   National Institute of Occupational Safety and Health

N.O.S.   Not Otherwise Specified

NRC   National Response Center

ORM   Other regulated material

OSHA   Occupational Safety and Health Administration

P   Polymerization hazard

PASS   Personal Alert Safety System

PEL   Permissible exposure limit

PG   Packing group

PID   Photo-ionization detector

PIH   Poison inhalation hazard

PPE   Personal protective equipment

RAID   Rapid Assessment and Initial Detection teams (National Guard)

RDD   Radiological dispersion device

REL   Recommended exposure level

R&I   Recognition and identification

RQ   Reportable quantity

SADT   Self-accelerating decomposition temperature

SAR   Supplied air respirators

SARA   Superfund Amendments and Reauthorization Act

SBCCOM   Soldiers Biological and Chemical Command

SCA   Surface contaminated articles

SCBA   Self-contained breathing apparatus

SERC   State Emergency Response Committee

SETIQ   Mexican Emergency Transportation System

SLUDGEM   Salivation, lacrimation, urination, defecation, gastrointestinal, emesis, miosis

STCC   Standard Transportation Commodity Code

STEL   Short term exposure limit

TC   Transportation Canada

TEU   Technical Escort Unit (U.S. Army)

TLV Threshold limit value

TOG Turnout gear

TRACEM Thermal, radiation, asphyxiation, chemical, etiological, mechanical hazard

TWA Time weighted average

UEL Upper explosive limit

UN United Nations identification number

USAR Urban Search and Rescue team (FEMA)

UST Underground storage tank

VTR Violent tank rupture

WMD Weapons of mass destruction

INDEX